BRITANNICA

Mathematics in Context

Insights into Data

Britannica

ENCYCLOPÆDIA BRITANNICA EDUCATIONAL CORPORATION

Mathematics in Context is a comprehensive curriculum for the middle grades. It was developed in collaboration with the Wisconsin Center for Education Research, School of Education, University of Wisconsin–Madison and the Freudenthal Institute at the University of Utrecht, The Netherlands, with the support of National Science Foundation Grant No. 9054928.

National Science Foundation

Opinions expressed are those of the authors
and not necessarily those of the Foundation

ISBN 0-7826-1555-4
1 2 3 4 5 WK 02 01 00 99 98

The *Mathematics in Context* Development Team

Mathematics in Context is a comprehensive curriculum for the middle grades. The National Science Foundation funded the National Center for Research in Mathematical Sciences Education at the University of Wisconsin–Madison to develop and field-test the materials from 1991 through 1996. The Freudenthal Institute at the University of Utrecht in The Netherlands, as a subcontractor, collaborated with the University of Wisconsin–Madison on the development of the curriculum.

The initial version of *Insights into Data* was developed by Monica Wijers and Jan de Lange. It was adapted for use in American schools by Mary C. Shafer and Gail Burrill.

National Center for Research in Mathematical Sciences Education Staff

Thomas A. Romberg
Director

Joan Daniels Pedro
Assistant to the Director

Gail Burrill
Coordinator
Field Test Materials

Margaret R. Meyer
Coordinator
Pilot Test Materials

Mary Ann Fix
Editorial Coordinator

Sherian Foster
Editorial Coordinator

James A. Middleton
Pilot Test Coordinator

Margaret A. Pligge
First Edition Coordinator

Project Staff

Jonathan Brendefur
Laura J. Brinker
James Browne
Jack Burrill
Rose Byrd
Peter Christiansen
Barbara Clarke
Doug Clarke
Beth R. Cole
Fae Dremock
Jasmina Milinkovic
Kay Schultz
Mary C. Shafer
Julia A. Shew
Aaron N. Simon
Marvin Smith
Stephanie Z. Smith
Mary S. Spence
Kathleen A. Steele

Freudenthal Institute Staff

Jan de Lange
Director

Els Feijs
Coordinator

Martin van Reeuwijk
Coordinator

Project Staff

Mieke Abels
Nina Boswinkel
Frans van Galen
Koeno Gravemeijer
Marja van den Heuvel-Panhuizen
Jan Auke de Jong
Vincent Jonker
Ronald Keijzer
Martin Kindt
Jansie Niehaus
Nanda Querelle
Anton Roodhardt
Leen Streefland
Adri Treffers
Monica Wijers
Astrid de Wild

Acknowledgments

Several school districts used and evaluated one or more versions of the materials: Ames Community School District, Ames, Iowa; Parkway School District, Chesterfield, Missouri; Stoughton Area School District, Stoughton, Wisconsin; Madison Metropolitan School District, Madison, Wisconsin; Milwaukee Public Schools, Milwaukee, Wisconsin; and Dodgeville School District, Dodgeville, Wisconsin. Two sites were involved in staff developments as well as formative evaluation of materials: Culver City, California, and Memphis, Tennessee. Two sites were developed through partnership with Encyclopædia Britannica Educational Corporation: Miami, Florida, and Puerto Rico. University Partnerships were developed with mathematics educators who worked with preservice teachers to familiarize them with the curriculum and to obtain their advice on the curriculum materials. The materials were also used at several other schools throughout the United States.

We at Encyclopædia Britannica Educational Corporation extend our thanks to all who had a part in making this program a success. Some of the participants instrumental in the program's development are as follows:

Allapattah Middle School
Miami, Florida
Nemtalla (Nikolai) Barakat

Ames Middle School
Ames, Iowa
Kathleen Coe
Judd Freeman
Gary W. Schnieder
Ronald H. Stromen
Lyn Terrill

Bellerive Elementary
Creve Coeur, Missouri
Judy Hetterscheidt
Donna Lohman
Gary Alan Nunn
Jakke Tchang

Brookline Public Schools
Brookline, Massachusetts
Rhonda K. Weinstein
Deborah Winkler

Cass Middle School
Milwaukee, Wisconsin
Tami Molenda
Kyle F. Witty

Central Middle School
Waukesha, Wisconsin
Nancy Reese

Craigmont Middle School
Memphis, Tennessee
Sharon G. Ritz
Mardest K. VanHooks

Crestwood Elementary
Madison, Wisconsin
Diane Hein
John Kalson

Culver City Middle School
Culver City, California
Marilyn Culbertson
Joel Evans
Joy Ellen Kitzmiller
Patricia R. O'Connor
Myrna Ann Perks, Ph.D.
David H. Sanchez
John Tobias
Kelley Wilcox

Cutler Ridge Middle School
Miami, Florida
Lorraine A. Valladares

Dodgeville Middle School
Dodgeville, Wisconsin
Jacqueline A. Kamps
Carol Wolf

Edwards Elementary
Ames, Iowa
Diana Schmidt

Fox Prairie Elementary
Stoughton, Wisconsin
Tony Hjelle

Grahamwood Elementary
Memphis, Tennessee
M. Lynn McGoff
Alberta Sullivan

Henry M. Flagler Elementary
Miami, Florida
Frances R. Harmon

Horning Middle School
Waukesha, Wisconsin
Connie J. Marose
Thomas F. Clark

Huegel Elementary
Madison, Wisconsin
Nancy Brill
Teri Hedges
Carol Murphy

Hutchison Middle School
Memphis, Tennessee
Maria M. Burke
Vicki Fisher
Nancy D. Robinson

Idlewild Elementary
Memphis, Tennessee
Linda Eller

Jefferson Elementary
Santa Ana, California
Lydia Romero-Cruz

Jefferson Middle School
Madison, Wisconsin
Jane A. Beebe
Catherine Buege
Linda Grimmer
John Grueneberg
Nancy Howard
Annette Porter
Stephen H. Sprague
Dan Takkunen
Michael J. Vena

Jesus Sanabria Cruz School
Yabucoa, Puerto Rico
Andreíta Santiago Serrano

John Muir Elementary School
Madison, Wisconsin
Julie D'Onofrio
Jane M. Allen-Jauch
Kent Wells

Kegonsa Elementary
Stoughton, Wisconsin
Mary Buchholz
Louisa Havlik
Joan Olsen
Dominic Weisse

Linwood Howe Elementary
Culver City, California
Sandra Checel
Ellen Thireos

Mitchell Elementary
Ames, Iowa
Henry Gray
Matt Ludwig

New School of Northern Virginia
Fairfax, Virginia
Denise Jones

Northwood Elementary
Ames, Iowa
Eleanor M. Thomas

Orchard Ridge Elementary
Madison, Wisconsin
Mary Paquette
Carrie Valentine

Parkway West Middle School
Chesterfield, Missouri
Elissa Aiken
Ann Brenner
Gail R. Smith

Ridgeway Elementary
Ridgeway, Wisconsin
Lois Powell
Florence M. Wasley

Roosevelt Elementary
Ames, Iowa
Linda A. Carver

Roosevelt Middle
Milwaukee, Wisconsin
Sandra Simmons

Ross Elementary
Creve Coeur, Missouri
Annette Isselhard
Sheldon B. Korklan
Victoria Linn
Kathy Stamer

St. Joseph's School
Dodgeville, Wisconsin
Rita Van Dyck
Sharon Wimer

St. Maarten Academy
St. Peters, St. Maarten, NA
Shareed Hussain

Sarah Scott Middle School
Milwaukee, Wisconsin
Kevin Haddon

Sawyer Elementary
Ames, Iowa
Karen Bush Hoiberg

Sennett Middle School
Madison, Wisconsin
Brenda Abitz
Lois Bell
Shawn M. Jacobs

Sholes Middle School
Milwaukee, Wisconsin
Chris Gardner
Ken Haddon

Stephens Elementary
Madison, Wisconsin
Katherine Hogan
Shirley M. Steinbach
Kathleen H. Vegter

Stoughton Middle School
Stoughton, Wisconsin
Sally Bertelson
Polly Goepfert
Jacqueline M. Harris
Penny Vodak

Toki Middle School
Madison, Wisconsin
Gail J. Anderson
Vicky Grice
Mary M. Ihlenfeldt
Steve Jernegan
Jim Leidel
Theresa Loehr
Maryann Stephenson
Barbara Takkunen
Carol Welsch

Trowbridge Elementary
Milwaukee, Wisconsin
Jacqueline A. Nowak

W. R. Thomas Middle School
Miami, Florida
Michael Paloger

Wooddale Elementary Middle School
Memphis, Tennessee
Velma Quinn Hodges
Jacqueline Marie Hunt

Yahara Elementary
Stoughton, Wisconsin
Mary Bennett
Kevin Wright

Site Coordinators

Mary L. Delagardelle—Ames Community Schools, Ames, Iowa

Dr. Hector Hirigoyen—Miami, Florida

Audrey Jackson—Parkway School District, Chesterfield, Missouri

Jorge M. López—Puerto Rico

Susan Militello—Memphis, Tennessee

Carol Pudlin—Culver City, California

Reviewers and Consultants

Michael N. Bleicher
Professor of Mathematics
University of Wisconsin–Madison
Madison, WI

Diane J. Briars
Mathematics Specialist
Pittsburgh Public Schools
Pittsburgh, PA

Donald Chambers
Director of Dissemination
University of Wisconsin–Madison
Madison, WI

Don W. Collins
Assistant Professor of Mathematics Education
Western Kentucky University
Bowling Green, KY

Joan Elder
Mathematics Consultant
Los Angeles Unified School District
Los Angeles, CA

Elizabeth Fennema
Professor of Curriculum and Instruction
University of Wisconsin–Madison
Madison, WI

Nancy N. Gates
University of Memphis
Memphis, TN

Jane Donnelly Gawronski
Superintendent
Escondido Union High School
Escondido, CA

M. Elizabeth Graue
Assistant Professor of Curriculum and Instruction
University of Wisconsin–Madison
Madison, WI

Jodean E. Grunow
Consultant
Wisconsin Department of Public Instruction
Madison, WI

John G. Harvey
Professor of Mathematics and Curriculum & Instruction
University of Wisconsin–Madison
Madison, WI

Simon Hellerstein
Professor of Mathematics
University of Wisconsin–Madison
Madison, WI

Elaine J. Hutchinson
Senior Lecturer
University of Wisconsin–Stevens Point
Stevens Point, WI

Richard A. Johnson
Professor of Statistics
University of Wisconsin–Madison
Madison, WI

James J. Kaput
Professor of Mathematics
University of Massachusetts–Dartmouth
Dartmouth, MA

Richard Lehrer
Professor of Educational Psychology
University of Wisconsin–Madison
Madison, WI

Richard Lesh
Professor of Mathematics
University of Massachusetts–Dartmouth
Dartmouth, MA

Mary M. Lindquist
Callaway Professor of Mathematics Education
Columbus College
Columbus, GA

Baudilio (Bob) Mora
Coordinator of Mathematics & Instructional Technology
Carrollton-Farmers Branch Independent School District
Carrollton, TX

Paul Trafton
Professor of Mathematics
University of Northern Iowa
Cedar Falls, IA

Norman L. Webb
Research Scientist
University of Wisconsin–Madison
Madison, WI

Paul H. Williams
Professor of Plant Pathology
University of Wisconsin–Madison
Madison, WI

Linda Dager Wilson
Assistant Professor
University of Delaware
Newark, DE

Robert L. Wilson
Professor of Mathematics
University of Wisconsin–Madison
Madison, WI

TABLE OF CONTENTS

BRITANNICA Mathematics in Context

Dear Teacher,

Welcome! *Mathematics in Context* is designed to reflect the National Council of Teachers of Mathematics Standards for School Mathematics and to ground mathematical content in a variety of real-world contexts. Rather than relying on you to explain and demonstrate generalized definitions, rules, or algorithms, students investigate questions directly related to a particular context and construct mathematical understanding and meaning from that context.

The curriculum encompasses 10 units per grade level. *Insights into Data* is designed to be the first unit in the statistics strand for grade 8/9, but it also lends itself to independent use—to introduce students to experiences that will enrich their understanding of gathering, interpreting, and presenting data, and how to critically analyze data sets and data representations.

In addition to the Teacher Guide and Student Books, *Mathematics in Context* offers the following components that will inform and support your teaching:

- *Teacher Resource and Implementation Guide,* which provides an overview of the complete system, including program implementation, philosophy, and rationale
- *Number Tools,* Volumes 1 and 2, which are a series of blackline masters that serve as review sheets or practice pages involving number issues/basic skills
- *News in Numbers,* which is a set of additional activities that can be inserted between or within other units; it includes a number of measurement problems that require estimation.

Thank you for choosing *Mathematics in Context.* We wish you success and inspiration!

Sincerely,

The Mathematics in Context Development Team

Overview

24 cm
BRITANNICA
Mathematics in Context
Year
World Population in Billions
FROG NEWTON
Portland
Eugene
Bend
Grants Pass
Klamath Falls
FAVORITE SPORTS

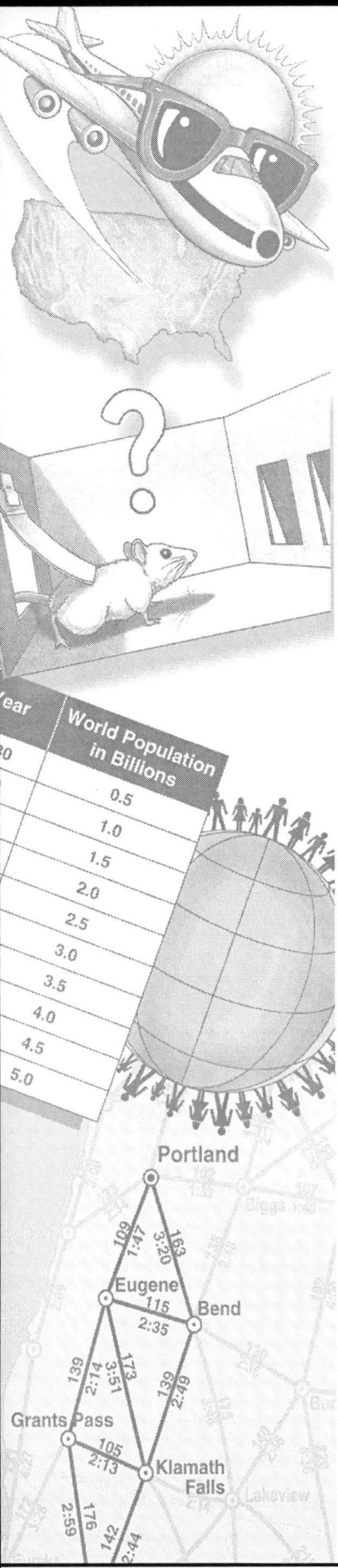

How to Use This Book

This unit is one of 40 for the middle grades. Each unit can be used independently; however, the 40 units are designed to make up a complete, connected curriculum (10 units per grade level). There is a Student Book and a Teacher Guide for each unit.

Each Teacher Guide comprises elements that assist the teacher in the presentation of concepts and in understanding the general direction of the unit and the program as a whole. Becoming familiar with this structure will make using the units easier.

Each Teacher Guide consists of six basic parts:

- Overview
- Student Materials and Teaching Notes
- Assessment Activities and Solutions
- Glossary
- Blackline Masters
- Try This! Solutions

Overview

Before beginning this unit, read the Overview in order to understand the purpose of the unit and to develop strategies for facilitating instruction. The Overview provides helpful information about the unit's focus, pacing, goals, and assessment, as well as explanations about how the unit fits with the rest of the *Mathematics in Context* curriculum.

Student Materials and Teaching Notes

This Teacher Guide contains all of the student pages (except the Try This! activities), each of which faces a page of solutions, samples of students' work, and hints and comments about how to facilitate instruction. Note: Solutions for the Try This! activities can be found at the back of this Teacher Guide.

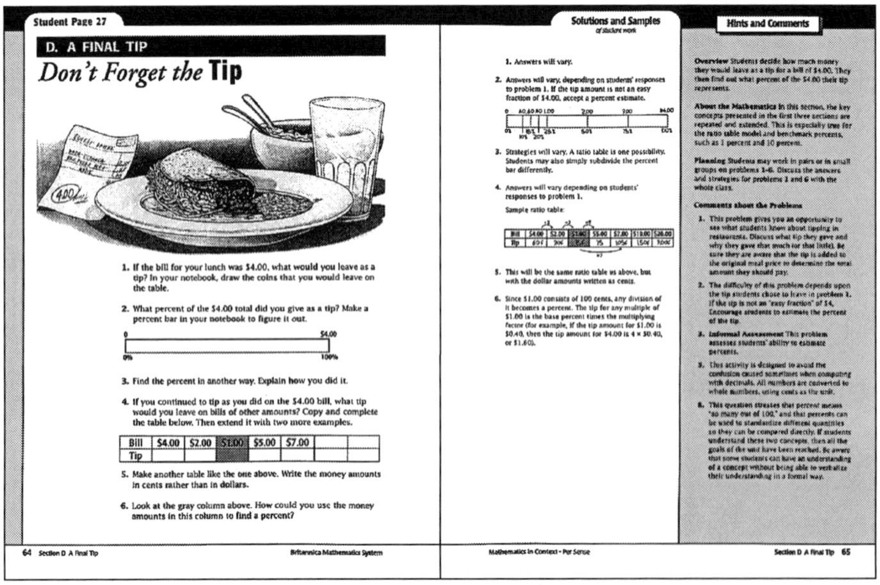
Student Page 27

D. A FINAL TIP

Don't Forget the **Tip**

1. If the bill for your lunch was $4.00, what would you leave as a tip? In your notebook, draw the coins that you would leave on the table.
2. What percent of the $4.00 total did you give as a tip? Make a percent bar in your notebook to figure it out.
3. Find the percent in another way. Explain how you did it.
4. If you continued to tip as you did on the $4.00 bill, what tip would you leave on bills of other amounts? Copy and complete the table below. Then extend it with two more examples.

Bill	$4.00	$2.00	$1.00	$5.00	$7.00		
Tip							

5. Make another table like the one above. Write the money amounts in cents rather than in dollars.
6. Look at the gray column above. How could you use the money amounts in this column to find a percent?

64 Section D A Final Tip — Britannica Mathematics System

Solutions and Samples
of student work

1. Answers will vary.
2. Answers will vary, depending on students' responses to problem 1. If the tip amount is not an easy fraction of $4.00, accept a percent estimate.
3. Strategies will vary. A ratio table is one possibility. Students may also simply subdivide the percent bar differently.
4. Answers will vary depending on students' responses to problem 1.

Sample ratio table:

Bill	$4.00	$2.00	$1.00	$5.00	$7.00	$10.00	$20.00
Tip	60¢	30¢	[illegible]	75	105¢	150¢	300¢

5. This will be the same ratio table as above, but with the dollar amounts written as cents.
6. Since $1.00 consists of 100 cents, any division of it becomes a percent. The tip for any multiple of $1.00 is the base percent times the multiplying factor (for example, if the tip amount for $1.00 is $0.40, then the tip amount for $4.00 is 4 × $0.40, or $1.60).

Mathematics in Context • Per Sense

Hints and Comments

Overview Students decide how much money they would leave as a tip for a bill of $4.00. They then find out what percent of the $4.00 their tip represents.

About the Mathematics In this section, the key concepts presented in the first three sections are repeated and extended. This is especially true for the ratio table model and benchmark percents, such as 1 percent and 10 percent.

Planning Students may work in pairs or in small groups on problems 1–6. Discuss the answers and strategies for problems 1 and 6 with the whole class.

Comments about the Problems

1. This problem gives you an opportunity to see what students know about tipping in restaurants. Discuss what tip they gave and why they gave that much (or that little). Be sure they are aware that the tip is added to the original meal price to determine the total amount they should pay.
2. The difficulty of this problem depends upon the tip students chose to leave in problem 1. If the tip is not an "easy fraction" of $4, Encourage students to estimate the percent of the tip.
3. **Informal Assessment** This problem assesses students' ability to estimate percents.
5. This activity is designed to avoid the confusion caused sometimes when computing with decimals. All numbers are converted to whole numbers, using cents as the unit.
6. This question stresses that percent means "so many out of 100," and that percents can be used to standardize different quantities so they can be compared directly. If students understand these two concepts, then all the goals of the unit have been reached. Be aware that some students can have an understanding of a concept without being able to verbalize their understanding in a formal way.

Section D A Final Tip 65

Each section within the unit begins with a two-page spread that describes the work students do, the goals of the section, new vocabulary, and materials needed, as well as providing information about the mathematics in the section and ideas for pacing, planning instruction, homework, and assessment.

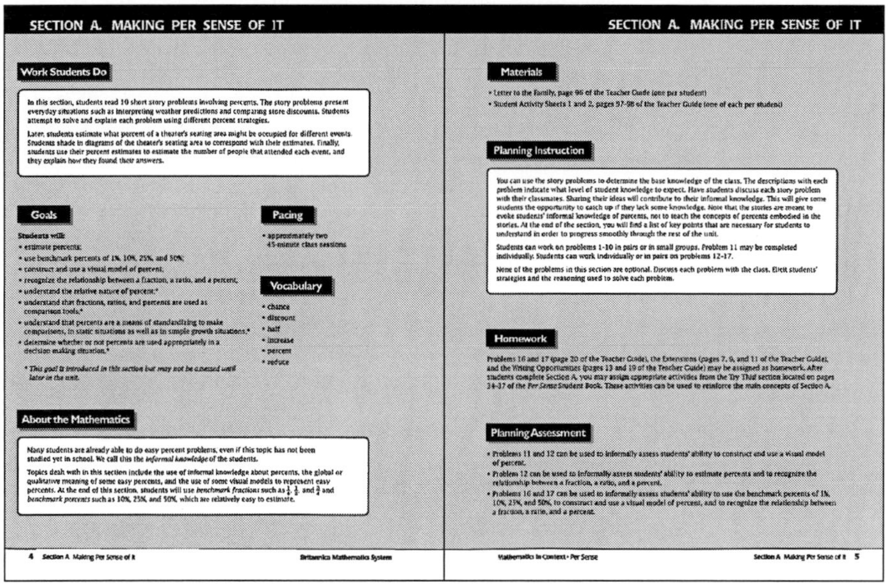
SECTION A. MAKING PER SENSE OF IT

Work Students Do

In this section, students read 10 short story problems involving percents. The story problems present everyday situations such as interpreting weather predictions and comparing store discounts. Students attempt to solve and explain each problem using different percent strategies.

Later, students estimate what percent of a theater's seating area might be occupied for different events. Students shade in diagrams of the theater's seating area to correspond with their estimates. Finally, students use their percent estimates to estimate the number of people that attended each event, and they explain how they found their answers.

Goals

Students will:
- estimate percents;
- use benchmark percents of 1%, 10%, 25%, and 50%;
- construct and use a visual model of percent;
- recognize the relationship between a fraction, a ratio, and a percent;
- understand the relative nature of percent;*
- understand that fractions, ratios, and percents are used as comparison tools;*
- understand that percents are a means of standardizing to make comparisons, in static situations as well as in simple growth situations;*
- determine whether or not percents are used appropriately in a decision making situation.*

** This goal is introduced in this section but may not be assessed until later in the unit.*

Pacing

- approximately two 45-minute class sessions

Vocabulary

- chance
- discount
- half
- increase
- percent
- reduce

About the Mathematics

Many students are already able to do easy percent problems, even if this topic has not been studied yet in school. We call this the *informal knowledge* of the students.

Topics dealt with in this section include the use of informal knowledge about percents, the global or qualitative meaning of some easy percents, and the use of some visual models to represent easy percents. At the end of this section, students will use *benchmark fractions* such as $\frac{1}{4}$, $\frac{1}{2}$, and $\frac{3}{4}$ and *benchmark percents* such as 10%, 25%, and 50%, which are relatively easy to estimate.

4 Section A Making Per Sense of It Britannica Mathematics System

SECTION A. MAKING PER SENSE OF IT

Materials

- Letter to the Family, page 96 of the Teacher Guide (one per student)
- Student Activity Sheets 1 and 2, pages 97–98 of the Teacher Guide (one of each per student)

Planning Instruction

You can use the story problems to determine the base knowledge of the class. The descriptions with each problem indicate what level of student knowledge to expect. Have students discuss each story problem with their classmates. Sharing their ideas will contribute to their informal knowledge. This will give some students the opportunity to catch up if they lack some knowledge. Note that the stories are meant to evoke students' informal knowledge of percents, not to teach the concepts of percents embodied in the stories. At the end of the section, you will find a list of key points that are necessary for students to understand in order to progress smoothly through the rest of the unit.

Students can work on problems 1-10 in pairs or in small groups. Problem 11 may be completed individually. Students can work individually or in pairs on problems 12-17.

None of the problems in this section are optional. Discuss each problem with the class. Elicit students' strategies and the reasoning used to solve each problem.

Homework

Problems 16 and 17 (page 20 of the Teacher Guide), the Extensions (pages 7, 9, and 11 of the Teacher Guide), and the Writing Opportunities (pages 13 and 19 of the Teacher Guide) may be assigned as homework. After students complete Section A, you may assign appropriate activities from the Try This! section located on pages 34-37 of the *Per Sense* Student Book. These activities can be used to reinforce the main concepts of Section A.

Planning Assessment

- Problems 11 and 12 can be used to informally assess students' ability to construct and use a visual model of percent.
- Problem 12 can be used to informally assess students' ability to estimate percents and to recognize the relationship between a fraction, a ratio, and a percent.
- Problems 16 and 17 can be used to informally assess students' ability to use the benchmark percents of 1%, 10%, 25%, and 50%, to construct and use a visual model of percent, and to recognize the relationship between a fraction, a ratio, and a percent.

Mathematics in Context • Per Sense Section A Making Per Sense of It 5

Assessment Activities and Solutions

Information about assessment can be found in several places in this Teacher Guide. General information about assessment is given in the Overview; informal assessment opportunities are identified on the teacher pages that face each student page; and the Assessment Activities section of this guide provides formal assessment opportunities.

Glossary

The Glossary defines all vocabulary words listed on the Section Opener pages. It includes the mathematical terms that may be new to students, as well as words associated with the contexts introduced in the unit. (Note: The Student Book does not have a glossary. This allows students to construct their own definitions, based on their personal experiences with the unit activities.)

Blackline Masters

At the back of this Teacher Guide are blackline masters for photocopying. The blackline masters include a letter to families (to be sent home with students before beginning the unit), several student activity sheets, and assessment masters.

Try This! Solutions

Also included in the back of this Teacher Guide are the solutions to several Try This! activities—one related to each section of the unit—that can be used to reinforce the unit's main concepts. The Try This! activities are located in the back of the Student Book.

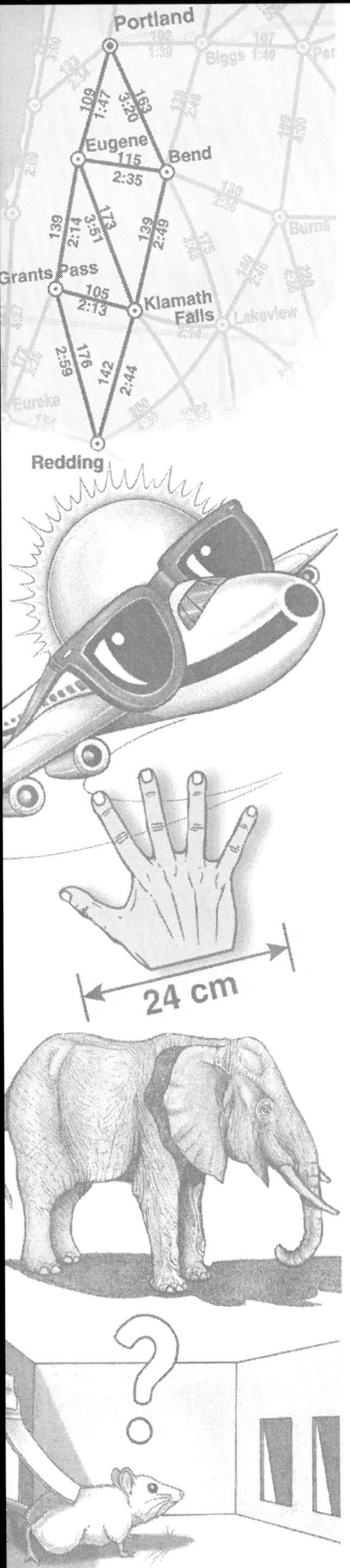

Unit Focus

The unit *Insights into Data* emphasizes an awareness of the questions that should be asked when analyzing data sets and representations of data, as well as when making conjectures and conclusions based on data. Students will use scatter plots, box plots, histograms, and numerical descriptions of data to investigate relationships in data sets. They will learn about the sampling process and possible sources of bias. Students will draw lines in scatter plots that summarize the data in order to predict, describe, check, or compare relationships among variables. Students also design and conduct a survey about an issue of importance to them. Connections are made throughout this unit to algebra and to geometry and measurement (area and volume).

Mathematical Content

- understanding graphical representations of data (scatter plots, box plots, number line plots, stem-and-leaf plots, and histograms)
- understanding random sampling
- choosing representative samples of a population
- describing data with statistical measures (mean, median, and mode)
- recognizing possible bias in survey results
- identifying the misrepresentation of data
- describing a correlation
- drawing straight lines (including the equation) that summarize data
- drawing conclusions based on data and representations of data
- collecting and representing data through survey, experiment, and simulation

Prior Knowledge

This unit assumes students can do the following with understanding:

- add, subtract, multiply, and divide rational numbers in all representations (decimals, percents, and fractions)
- read, interpret, make, and use graphical representations (scatter plots, box plots, number line plots, stem-and-leaf plots, and histograms) as introduced and used in the units *Picturing Numbers, Dealing with Data,* and *Statistics and the Environment*
- understand and use statistical measures (mean, median, and mode) as introduced and used in the units *Picturing Numbers, Dealing with Data,* and *Statistics and the Environment*
- perform random sampling as introduced in the unit *Dealing with Data*
- make and use the equations of a straight line (slope and y-intercept) as introduced in the unit *Graphing Equations*
- use metric measurements
- use rules to compute area and volume (as introduced in *Reallotment* and *Cereal Numbers*)

This unit should be taught after the statistics and probability units *Picturing Numbers* and *Dealing with Data,* the algebra unit *Graphing Equations,* and the geometry unit *Reallotment.*

Planning and Preparation

Pacing: 15–17 days

Section	Work Students Do	Pacing*	Materials
A. Patterns in Data	■ conduct an experiment and collect data on the growth of bean sprouts ■ interpret scatter plots and describe patterns in them	3 days	■ Letter to the Family (one per student) ■ Student Activity Sheet 1 (one per group) ■ Student Activity Sheet 2 (one per student) ■ See page 5 of the Teacher Guide for a complete list of materials and quantities needed.
B. Selecting Samples	■ investigate sampling processes ■ interpret graphical representations of data ■ perform random sampling and recognize possible bias in survey results ■ draw conclusions from data	3 days	■ Student Activity Sheets 3 and 4 (one per student) ■ See page 27 of the Teacher Guide for a complete list of materials and quantities needed.
C. Interpreting Graphs	■ interpret graphical representations of data ■ investigate the misrepresentation of data ■ recognize possible bias in survey results ■ draw conclusions from data	2–3 days	■ Student Activity Sheet 5 (one per group of students) ■ See page 49 of the Teacher Guide for a complete list of materials and quantities needed.
D. Using Plant Growth Data	■ select and use appropriate graphical representations to make comparisons ■ interpret graphical representations of data	3 days	■ See page 67 of the Teacher Guide for a complete list of materials and quantities needed.
E. Correlating Data	■ investigate relationships between two variables, using scatter plots ■ describe a correlation intuitively ■ draw conclusions based on data ■ represent data	1–2 days	■ Student Activity Sheet 2, optional (one per student)
F. Lines That Summarize Data	■ interpret and draw lines that summarize data ■ use correlation ■ use several techniques to draw best-fit lines ■ interpret graphical representations of data	3 days	■ Student Activity Sheets 6–9, 13, and 14 (one of each per group of students) ■ See page 103 of the Teacher Guide for a complete list of materials and quantities needed.

* One day is approximately equivalent to one 45-minute class session.

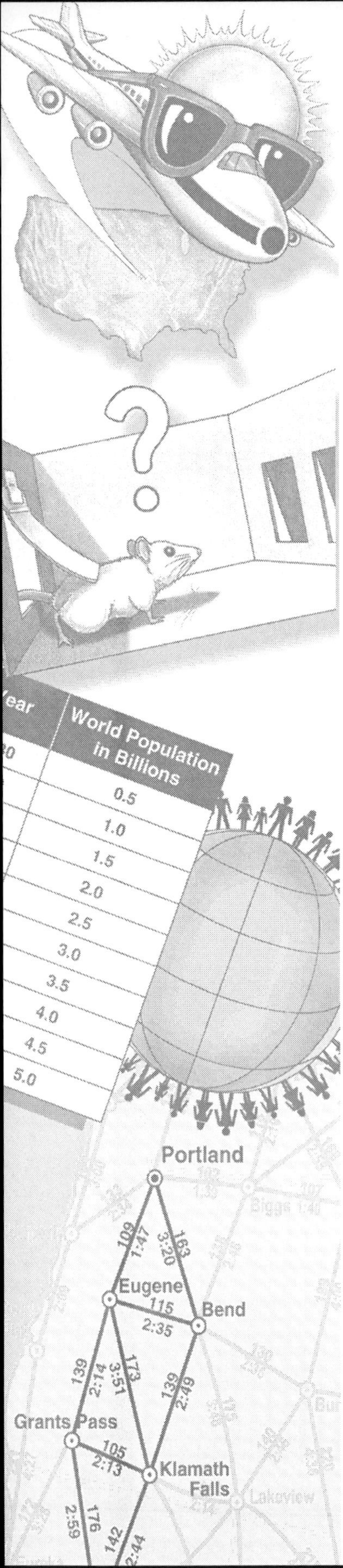

Preparation

In the *Teacher Resource and Implementation Guide* is an extensive description of the philosophy underlying both the content and the pedagogy of the *Mathematics in Context* curriculum. Suggestions for preparation are also given in the Hints and Comments columns of this Teacher Guide. You may want to consider the following:

- Work through the unit before teaching it. If possible, take on the role of the student and discuss your strategies with other teachers.
- Use the overhead projector for student demonstrations, particularly with overhead transparencies of the student activity sheets and any manipulatives used in the unit.
- Invite students to use drawings and examples to illustrate and clarify their answers.
- Allow students to work at different levels of sophistication. Some students may need concrete materials, while others can work at a more abstract level.
- Provide opportunities and support for students to share their strategies, which often differ. This allows students to take part in class discussions and introduces them to alternative ways to think about the mathematics in the unit.
- In some cases, it may be necessary to read the problems to students or to pair students to facilitate their understanding of the printed materials.
- A list of the materials needed for this unit is in the chart on page xiii.
- Try to follow the recommended pacing chart on page xiii. You can easily spend much more time on this unit than the number of class periods indicated. Bear in mind, however, that many of the topics introduced in this unit will be revisited and covered more thoroughly in other *Mathematics in Context* units.

Resources

For Teachers	For Students
 **Books and Magazines** • *Mathematics Assessment: Myths, Models, Good Questions, and Practical Suggestions*, edited by Jean Kerr Stenmark (Reston, Virginia: The National Council of Teachers of Mathematics, Inc., 1991) • *Exploring Data*, Landwehr, James, and Ann E. Watkins (Palo Alto, CA: Dale Seymour Publications, 1995)	 **Videos** Mathsphere Videos • *What's the Connection?* • *Compact Costing* • *Facts into Pictures* • *A Need for Data* • *An Exhausting Quest* • *A Value for Cleanliness* • *Polls Apart* (available from Encyclopædia Britannica)

Assessment

Planning Assessment

In keeping with the NCTM Assessment Standards, valid assessment should be based on evidence drawn from several sources. (See the full discussion of assessment philosophies in the *Teacher Resource and Implementation Guide.*) An assessment plan for this unit may draw from the following sources:

- Observations—look, listen, and record observable behavior.
- Interactive Responses—in a teacher-facilitated situation, note how students respond, clarify, revise, and extend their thinking.
- Products—look for the quality of thought evident in student projects, test answers, worksheet solutions, or writings.

These categories are not meant to be mutually exclusive. In fact, observation is a key part in assessing interactive responses and also key to understanding the end results of projects and writings.

Ongoing Assessment Opportunities

- **Problems within Sections**
  To evaluate ongoing progress, *Mathematics in Context* identifies informal assessment opportunities and the goals that these particular problems assess throughout the Teacher Guide. There are also indications as to what you might expect from your students.

- **Section Summary Questions**
  The summary questions at the end of each section are vehicles for informal assessment (see Teacher Guide pages 24, 46, 64, 82, 100, and 128).

End-of-Unit Assessment Opportunities

In the back of this Teacher Guide, there are three assessment activities that can be completed in three 45-minute class sessions. For a more detailed description of the assessment activities, see the Assessment Overview (Teacher Guide pages 130 and 131). Since only a selection of the goals of the unit are assessed in the end-of-unit assessments, you may want to use problems from the sections to assess the other goals.

You may also wish to design your own culminating project. In the Design-a-Test assessment activity, your students create a test that will tell you what they consider is important in the unit. For more assessment ideas, refer to the chart on pages xvi and xvii.

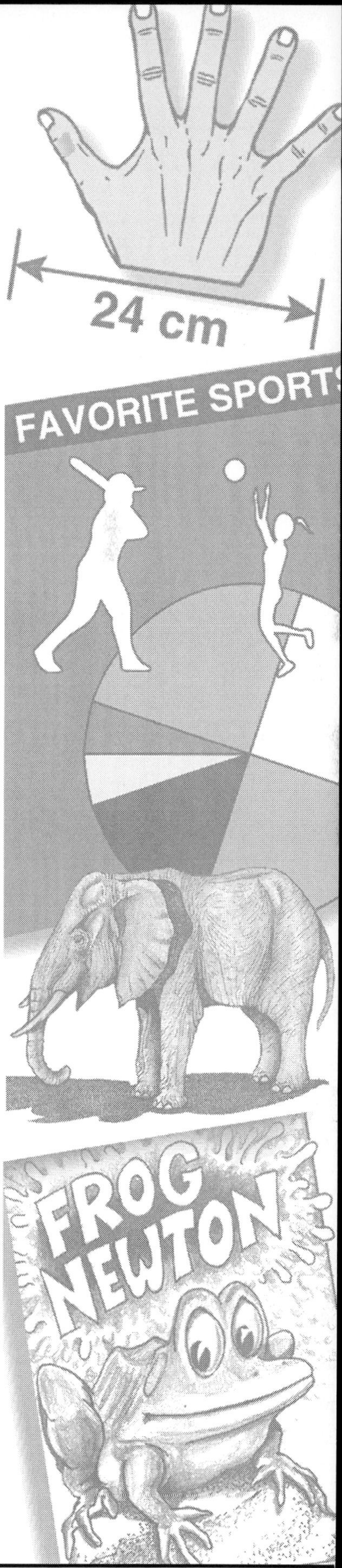

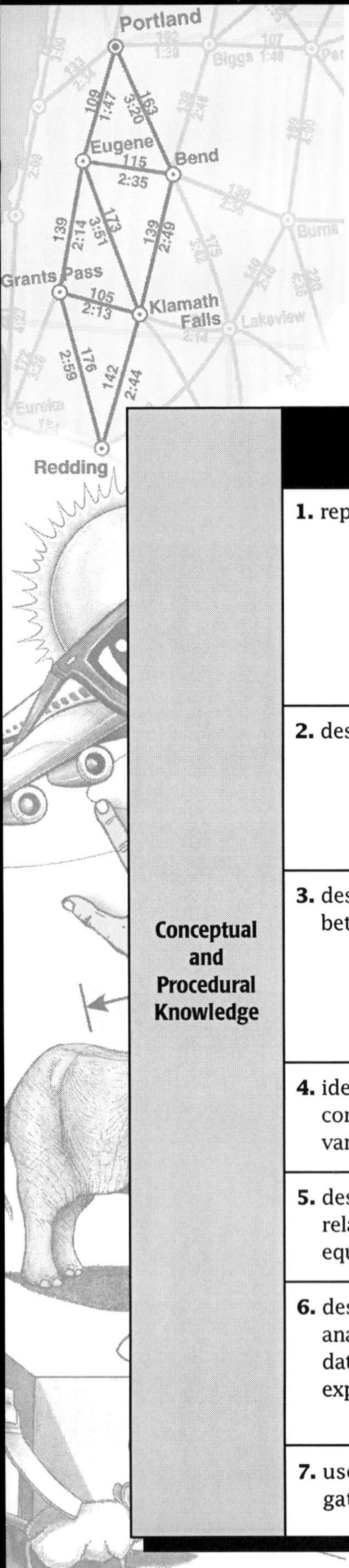

Goals and Assessment

In the *Mathematics in Context* curriculum, unit goals, categorized according to cognitive procedures, relate to the strand goals and to the NCTM Curriculum and Evaluation Standards. Additional information about these goals is found in the *Teacher Resource and Implementation Guide.* The *Mathematics in Context* curriculum is designed to help students develop their abilities so that they can perform with understanding in each of the categories listed below. It is important to note that the attainment of goals in one category is not a prerequisite to attaining those in another category. In fact, students should progress simultaneously toward several goals in different categories.

	Goal	Ongoing Assessment Opportunities		End-of-Unit Assessment Opportunities
Conceptual and Procedural Knowledge	**1.** represent data graphically	**Section B**	p. 36, #8, p. 44, #14, #16	Hats, p. 158 Design-a-Test, p. 157 Manatees and Power Boats, pp. 159–160
		Section C	p. 64, #14	
		Section D	p. 72, #3 p. 78, #5 p. 82, #8	
		Section F	p. 124, #21, #22	
	2. describe data numerically	**Section A**	p. 20, #12	Hats, p. 158 Design-a-Test, p. 157 Manatees and Power Boats, pp. 159–160
		Section B	p. 44, #14, #16	
		Section D	p. 78, #5 p. 82, #9	
		Section F	p. 124, #21, #22	
	3. describe the relationship between two variables	**Section C**	p. 58, #7	Hats, p. 158 Manatees and Power Boats, pp. 159–160 Design-a-Test, p. 157
		Section E	p. 94, #6 p. 100, #13, #14	
		Section F	p. 116, #13 p. 124, #21, #22, p. 126, #24	
	4. identify the degree of correlation between variables	**Section E**	p. 100, #13, #14	Hats, p. 158 Manatees and Power Boats, pp. 159–160
		Section F	p. 126, #24 p. 128, #26	
	5. describe a linear relationship with an equation of a straight line	**Section F**	p. 118, #14 p. 124, #21, #22	Design-a-Test, p. 157
	6. design, conduct and analyze ways of gathering data: surveys, simulations, experiments	**Section B**	p. 36, #8 p. 46, #19	Hats, p. 158 Manatees and Power Boats, pp. 159–160 Design-a-Test, p. 157
		Section C	p. 62, #12 p. 64, #14	
		Section D	p. 78, #5	
	7. use random samples in gathering data	**Section B**	pp. 40–42, #10	Design-a-Test, p. 157
		Section C	p. 62, #12	

	Goal	Ongoing Assessment Opportunities	End-of-Unit Assessment Opportunities
Reasoning, Communicating, Thinking, and Making Connections	**8.** analyze representations of data	**Section A** p. 14, #5 p. 20, #12 p. 22, #15 p. 24, #16 **Section C** p. 54, #5 p. 58, #7 p. 64, #13, #14 **Section D** p. 72, #3 p. 78, #5 **Section E** p. 88, #3 p. 90, #5 **Section F** p. 116, #13	Design-a-Test, p. 157
	9. draw conclusions based on given data sets and representations of data	**Section A** p. 20, #12 p. 22, #15 **Section B** p. 36, #8 p. 44, #14, #16 **Section C** p. 64, #14 **Section D** p. 78, #5 **Section E** p. 94, #6	Design-a-Test, p. 157 Hats, p. 158 Manatees and Power Boats, pp. 159–160

	Goal	Ongoing Assessment Opportunities	End-of-Unit Assessment Opportunities
Modeling, Nonroutine Problem-Solving, Critically Analyzing, and Generalizing	**10.** recognize possible bias in sample surveys	**Section B** pp. 40–42, #10 p. 46, #18, #19 **Section C** p. 58, #7 p. 62, #12	Design-a-Test, p. 157
	11. determine whether representations of data (numerical and visual) are appropriate	**Section B** pp. 40–42, #10 **Section C** p. 54, #5 p. 58, #7 p. 64, #13, #14 **Section D** p. 82, #8, #9	Design-a-Test, p. 157 Hats, p. 158 Manatees and Power Boats, pp. 159–160
	12. become aware of the questions that should be asked when analyzing data sets and representations of data	**Section A** p. 20, #12 p. 22, #15 **Section B** p. 36, #8 pp. 40–42, #10 p. 46, #18, #19 **Section C** p. 64, #13, #14 **Section D** p. 78, #5 p. 82, #9 **Section F** p. 126, #24 p. 127, Extension	Design-a-Test, p. 157 Hats, p. 158 Manatees and Power Boats, pp. 159–160

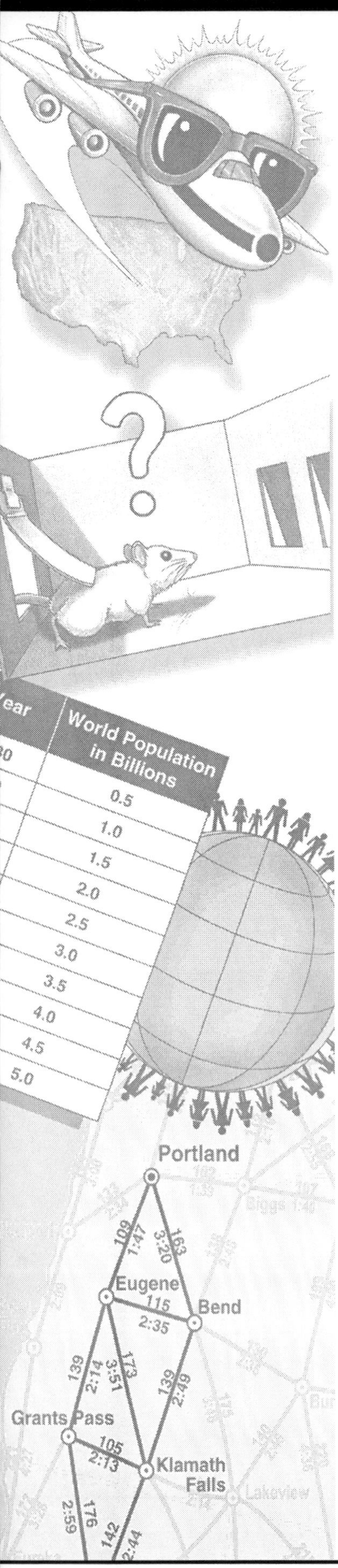

More about Assessment

Scoring and Analyzing Assessment Responses

Students may respond to assessment questions with various levels of mathematical sophistication and elaboration. Each student's response should be considered for the mathematics that it shows, and not judged on whether or not it includes an expected response. Responses to some of the assessment questions may be viewed as either correct or incorrect, but many answers will need flexible judgment by the teacher. Descriptive judgments related to specific goals and partial credit often provide more helpful feedback than percent scores.

Openly communicate your expectations to all students, and report achievement and progress for each student relative to those expectations. When scoring students' responses, try to think about how they are progressing toward the goals of the unit and the strand.

Student Portfolios

Generally, a portfolio is a collection of student-selected pieces that is representative of a student's work. A portfolio may include evaluative comments by you or by the student. See the *Teacher Resource and Implementation Guide* for more ideas on portfolio focus and use.

A comprehensive discussion about the contents, management, and evaluation of portfolios can be found in *Mathematics Assessment: Myths, Models, Good Questions, and Practical Suggestions,* pp. 35–48.

Student Self-Evaluation

Self-evaluation encourages students to reflect on their progress in learning mathematical concepts, their developing abilities to use mathematics, and their dispositions toward mathematics. The following examples illustrate ways to incorporate student self-evaluations as one component of your assessment plan.

- Ask students to comment, in writing, on each piece they have chosen for their portfolios and on the progress they see in the pieces overall.
- Give a writing assignment entitled "What I Know Now about [a math concept] and What I Think about It." This will give you information about each student's disposition toward mathematics as well as his or her knowledge.
- Interview individuals or small groups to elicit what they have learned, what they think is important, and why.

Suggestions for self-inventories can be found in *Mathematics Assessment: Myths, Models, Good Questions, and Practical Suggestions,* pp. 55–58.

Summary Discussion

Discuss specific lessons and activities in the unit—what the student learned from them and what the activities have in common. This can be done in whole-class discussion, in small groups, or in personal interviews.

Connections across the *Mathematics in Context* Curriculum

Insights into Data is a grade 8/9 unit in the statistics strand. The map below shows the complete *Mathematics in Context* curriculum for grade 8/9. It indicates where the unit fits in the statistics strand, as well as in the overall picture.

A detailed description of the units, the strands, and the connections in the *Mathematics in Context* curriculum can be found in the *Teacher Resource and Implementation Guide.*

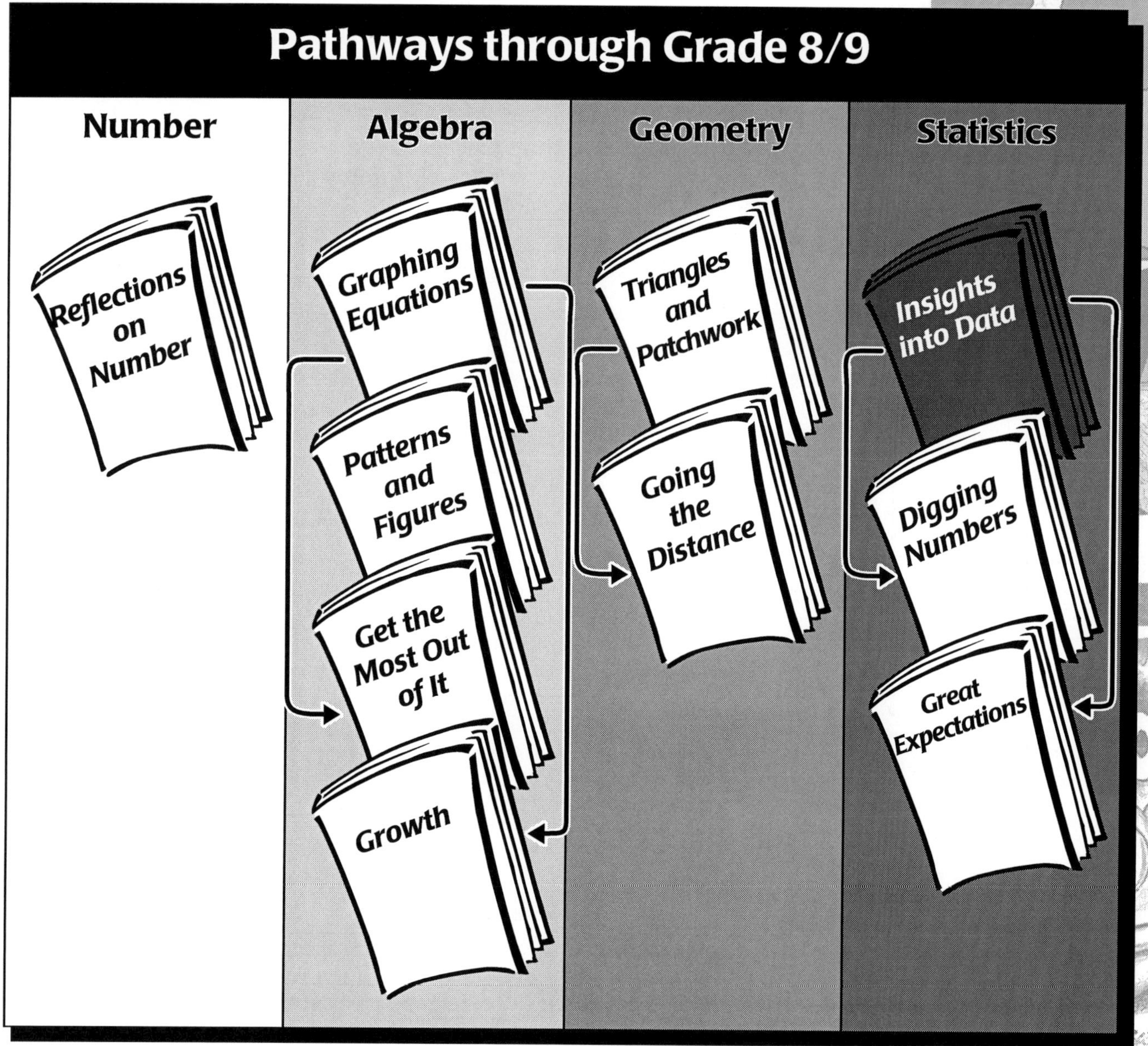

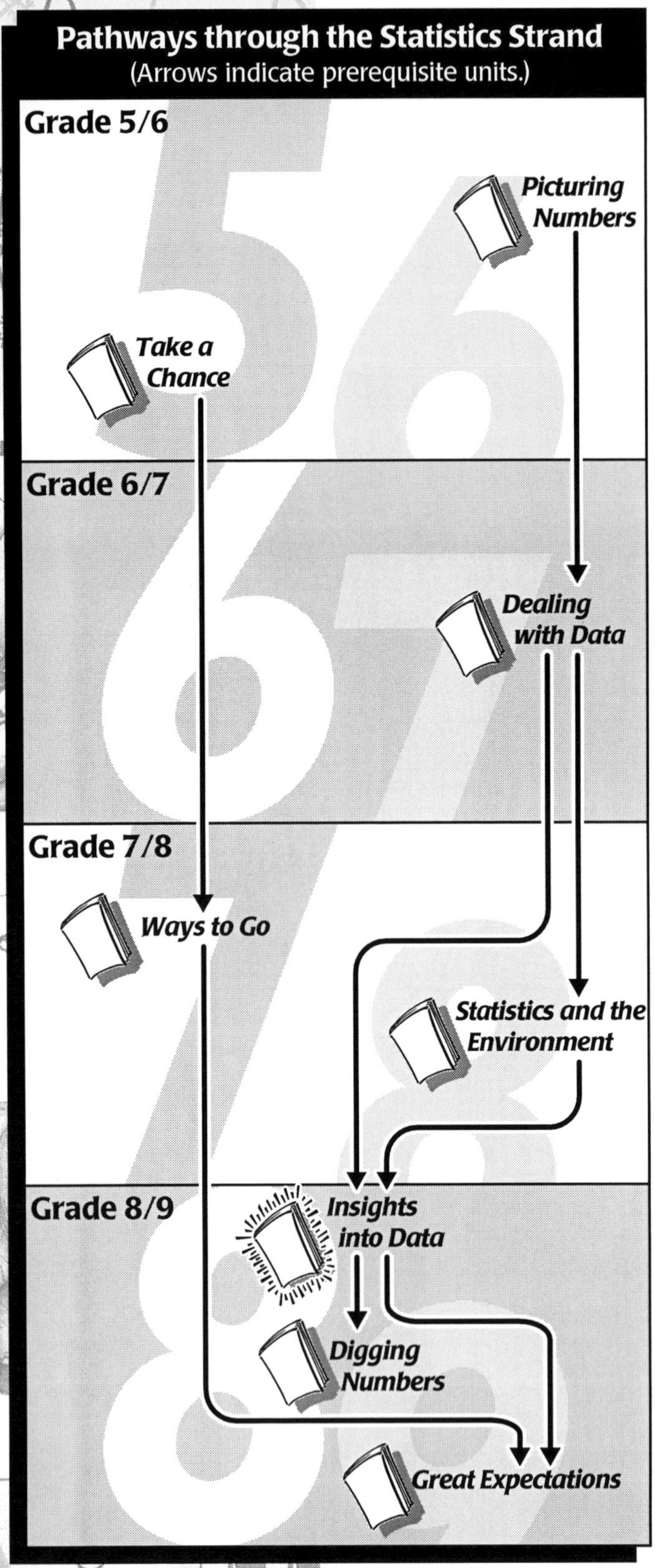

Connections within the Statistics Strand

On the left is a map of the statistics strand; this unit, *Insights into Data,* is highlighted.

The major themes in the statistics strand are dealing with data, developing an understanding of chance and probability, using probability in situations connected to statistics, and developing critical-thinking skills.

Collecting and displaying data is a central theme in the statistics units. Students should be able to describe data visually and numerically and make conclusions based on data and descriptions of data. In *Picturing Numbers,* students begin that process with an introduction to numerical descriptions of data and to the mean as a measure of center. The mean is revisited, and other summary measures (mode and median) are formally introduced, in the grade 6/7 unit *Dealing with Data.* Summary answers are revisited in the grade 7/8 unit *Statistics and the Environment* and in the grade 8/9 unit *Insights into Data.*

Graphical representations of data are found in all of the statistics units, especially in *Dealing with Data,* where histograms, scatter plots, stem-and-leaf plots, and box plots are introduced. Students use data and graphs to make conclusions and build arguments. These ideas also appear in units from other strands. For example, the mean is used in *Ratios and Rates* to build an understanding of rates, and line plots are the basis for the entire unit *Tracking Graphs* in grade 6/7. The equation of a straight line is introduced in the algebra unit *Graphing Equations* and used in *Insights into Data.*

The critical attitude that is developed in *Picturing Numbers* is extended in the units *Statistics and the Environment* and *Insights into Data.* The focus of *Insights into Data* is on the use of previously developed statistics (measures and graphs) and on developing a critical attitude toward using statistics.

Students have also been introduced to random samples in *Dealing with Data* and *Statistics and the Environment.* Sampling and drawing conclusions based on data from samples will be revisited in *Great Expectations.*

The Statistics Strand

Grade 5/6

Picturing Numbers
Creating and interpreting different kinds of graphs (bar and line graphs, pie charts, and pictographs), describing data in tabular and graphical forms, and using data as an aid to argument.

Take a Chance
Exploring the meanings of the terms *fair* and *chance;* computing chances in situations with few outcomes; using repeated trials to estimate chances; and using tree diagrams to represent two- or three-event situations.

Grade 7/8

Ways to Go
Reading and interpreting different kinds of maps, drawing graphs and networks, representing tables as graphs (and vice versa), solving problems with the help of connectivity graphs and tables, and solving simple combinatoric problems. (*Ways to Go* is also in the geometry strand.)

Statistics and the Environment
Collecting data; summarizing data with one-number summaries; reading and interpreting different kinds of maps, graphs, and tables; and using graphs and data as aids to argument.

Grade 6/7

Dealing with Data
Creating and interpreting scatter plots, box plots, stem-and-leaf diagrams, histograms, and number line plots; describing data in tabular and graphical forms; using one-number summaries (mean, median, mode, range, quartiles); and using data as an aid to argument.

Grade 8/9

Insights into Data
Analyzing information presented in a graph; gathering data using sampling techniques; recognizing bias in sample surveys; examining misleading representations of data; identifying weak, strong, positive, and negative correlations and no correlation; using lines to summarize the data in a scatter plot; and making and interpreting scatter plots and box plots.

Digging Numbers
Using the properties of height, diameter, and radius to determine whether or not various irregular shapes are similar; predicting length using graphs and formulas; exploring the relationship between three-dimensional shapes and drawings of them; and using length-to-width ratios to classify various objects. (*Digging Numbers* is also in the geometry strand.)

Great Expectations
Recognizing the variability in taking random samples and the role of chance in world events; understanding the effect of sample size on the variability in samples; finding the probability of independent and dependent events; using expected values to make decisions; using simulations to investigate probability; and counting all the possible arrangements of a set, with and without repetition.

Connections with Other *Mathematics in Context* Units

Insights into Data is the last unit in the *Mathematics in Context* curriculum that deals with statistics. The unit builds on students' understanding of measurement, area, and volume (from the geometry and measurement units) and on their knowledge of drawing straight lines and finding the corresponding equations, as introduced in the algebra strand.

The following mathematical topics that are included in the unit *Insights into Data* are introduced or further developed in other *Mathematics in Context* units.

Prerequisite Topics

Topic	Unit	Grade
bar graphs, line graphs	*Picturing Numbers*	5/6
	Tracking Graphs *	6/7
	Ups and Downs *	7/8
scatter plots, stem-and-leaf plots, box plots, number line plots	*Dealing with Data*	6/7
	Statistics and the Environment	7/8
collecting and describing data	*Picturing Numbers*	5/6
	Dealing with Data	6/7
	Statistics and the Environment	7/8
representativeness, sampling, population	*Dealing with Data*	6/7
	Statistics and the Environment	7/8
metric measurements	*Measure for Measure* **	5/6
	Made to Measure ***	6/7
mean, mode, and median	*Picturing Numbers*	5/6
	Dealing with Data	6/7
	Ratios and Rates **	6/7
area and volume	*Reallotment* ***	6/7
	Cereal Numbers **	7/8
	Made to Measure ***	6/7
slope and intercept	*Looking at an Angle* ***	7/8
	Ups and Downs *	7/8
	Graphing Equations *	8/9

Topics Revisited in Other Units

Topic	Unit	Grade
graphical representations	*Digging Numbers*	8/9
	Great Expectations	8/9
mean, median, mode, minimum, maximum, range	*Great Expectations*	8/9
graphing	*Get the Most Out of It* *	8/9
	Growth *	8/9
scatter plots	*Growth* *	8/9
	Digging Numbers	8/9

* These units in the algebra strand also help students make connections to ideas about statistics.
** These units in the number strand also help students make connections to ideas about statistics.
*** These units in the geometry strand also help students make connections to ideas about statistics.

BRITANNICA Mathematics in Context

Student Materials and Teaching Notes

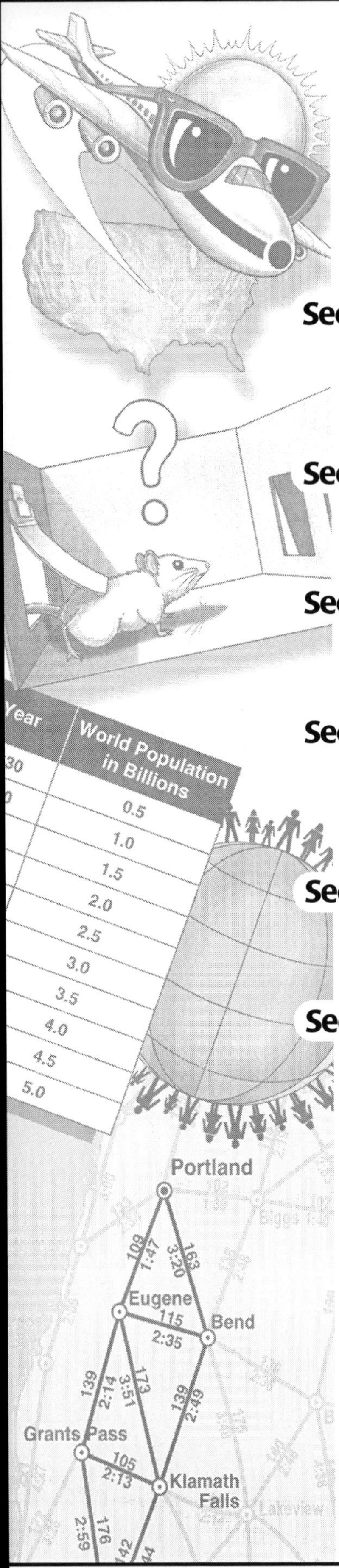

Student Book
Table of Contents

Dear Student,

Welcome to *Insights into Data.* Do you look at the graphs in newspapers to see if they make sense? Numbers and graphs are used as sources of data all around you: in sports, grades, sales and marketing, taxes, and car ratings.

Year	World Population (in billions)
1630	0.5
1820	1.0
1890	1.5
1930	2.0
1950	2.5
1960	3.0
1968	3.5
1975	4.0
1981	4.5
1987	5.0

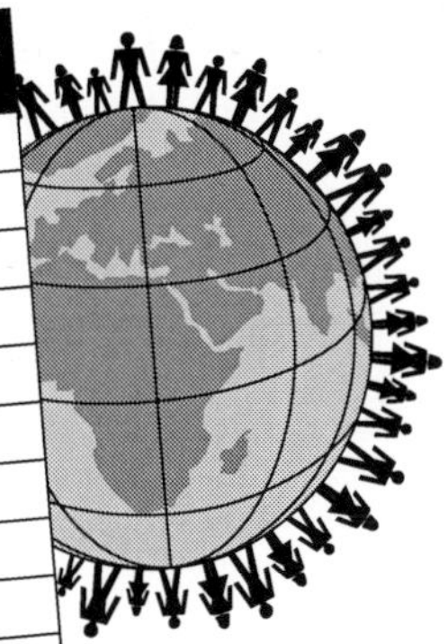

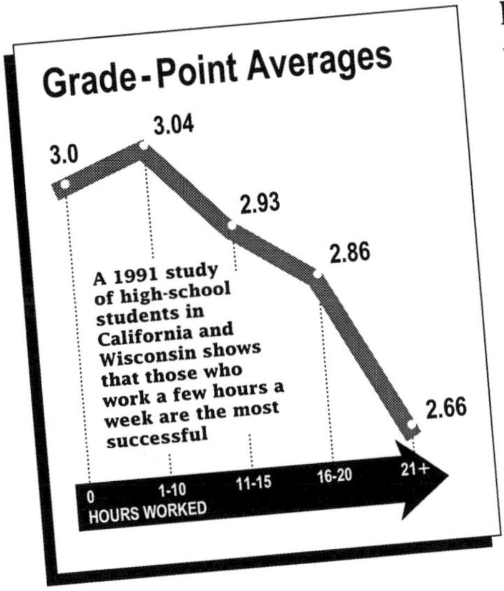

In this unit, you will learn how to use numbers and graphs to help you make decisions and draw conclusions. You will also study surveys and how they are conducted. You will grow mung beans in soda, salt water, and tap water to see which is the best solution for growing sprouts. You will even learn to use lines to help you investigate the relationship between two things, such as the height of a dinosaur and the length of its tail.

As you develop insights into data, keep your eye out for graphs in newspapers and magazines.

Sincerely,

The Mathematics in Context Development Team

Work Students Do

First, students plant bean sprouts in an experiment that simulates the effects of different chemicals on crop growth. They compare the variability in the lengths of the sprouts over seven days. Students then investigate statements about data represented in a graph. They explore a scatter plot and analyze underlying patterns in the data. Students find that it may be necessary to collect additional data in order for them to make appropriate inferences.

Goals

Students will:

- describe data numerically;
- describe the relationship between two variables;*
- design, conduct, and analyze ways of gathering data: surveys, simulations, and experiments;*
- analyze representations of data;
- draw conclusions based on given data sets and representations of data;
- become aware of the questions that should be asked when analyzing data sets and representations of data.

** These goals are introduced in this section and are assessed in other sections of the unit.*

Pacing

- approximately three 45-minute class sessions

Vocabulary

- cluster
- data point
- mean
- outlier
- population
- scatter plot

About the Mathematics

A scatter plot can be used to investigate the relationship between two quantitative variables. The variable you begin with (sometimes called the independent variable) is generally associated with the horizontal axis. The second variable (sometimes called the dependent variable) is generally placed on the vertical axis. There are often underlying patterns in scatter plots of paired data. Consider clusters of points, outliers, and trends. Identifying the data points as belonging to different categories can also help reveal patterns. For example, in the scatter plot on Student Book page 5, each data point can be circled in a color that indicates which region of the country that state is in. In this way, regional trends can be discovered. Critical review of representations of data may uncover more complex relationships than are initially noticed.

Materials

- Letter to the Family, page 142 of the Teacher Guide (one per student)
- Student Activity Sheet 1, page 143 of the Teacher Guide (one per group)
- Student Activity Sheet 2, page 144 of the Teacher Guide (one per student)
- See page 9 of the Teacher Guide for a complete list of materials and quantities needed for the bean sprout experiment.
- overhead transparency of Student Activity Sheet 2, page 15 of the Teacher Guide, optional (one per class)
- overhead projector, page 15 of the Teacher Guide, optional (one per class)
- maps of the United States, page 15 of the Teacher Guide, optional (one per student)
- blue and red markers, page 23 of the Teacher Guide, (one of each per student)

Planning Instruction

The unit starts with a growth experiment that gives students the opportunity to gather their own data that they will analyze later in the unit. You will need to prepare for the activity at least a day before starting the unit. Prepare the mung beans by soaking them overnight in tap water. Discard those that do not look like they will root. If mung beans are not available, other beans may be used.

You may want to begin this section with a discussion of what a population is. Students likely will know the term *population* as it refers to a group of people living in an area. Inform them that this definition can be extended to include any group of people or objects you want to gather information about. Scientists collect and analyze data about populations to understand more about them.

Students may work on problems 1 and 10–15 in small groups. You may want to discuss problems 2 and 3 as a class. Students may work on problems 4–9 individually or in small groups. They may work on problem 16 individually.

There are no optional problems in this section.

Homework

Problem 9 (page 18 of the Teacher Guide) may be assigned as homework. The Bringing Math Home activity (page 11 of the Teacher Guide), the Extensions (pages 13 and 15 of the Teacher Guide) and the Writing Opportunity (page 21 of the Teacher Guide) also can be assigned as homework. After students complete Section A, you may assign appropriate activities from the Try This! section, located on pages 58–62 of the *Insights into Data* Student Book. The Try This! activities reinforce the key mathematical concepts introduced in this section.

Planning Assessment

- Problems 5 and 16 can be used to informally assess students' ability to analyze representations of data.
- Problem 12 can be used to informally assess students' ability to describe data numerically and to analyze representations of data. It also assesses their ability to draw conclusions based on given data sets and representations of data and to become aware of the questions that should be asked when analyzing data sets and representations of data.
- Problem 15 can be used to informally assess students' ability to analyze representations of data and to draw conclusions based on given data sets and representations of data. It also assesses their ability to become aware of the questions that should be asked when analyzing data sets and representations of data.

A. PATTERNS IN DATA

Bean Sprout Experiment

Farmers are concerned about parasites damaging their crops. Chemical companies develop pesticides that kill the parasites, but they have to be careful that the chemicals do not harm the crops. In this section, you will collect and examine data, and you will study how graphs can help you reach conclusions about a data set. In the experiment that follows, you will examine the effect of different liquids on the growth of bean sprouts germinated from mung beans.

Mung Beans

Solutions and Samples
of student work

Hints and Comments

Overview Students are introduced to an activity in which they will investigate the effects of different solutions on the growth of bean sprouts. There are no problems on this page for students to solve.

About the Mathematics An experiment is introduced as a way to collect data. Other data-collection methods such as surveys and simulations get attention in other sections.

Planning You may want to read the text on Student Book page 1 as a class. Make sure students understand the text before moving on to the next page. In particular, students may have trouble with certain words, such as *parasite, pesticide,* and *germinate.*

The Design-a-Test assessment activity (see the Assessment Overview) will give students a unique opportunity to show what they have learned in this unit. If you choose to have students work on this assessment activity, you may want to explain the process now, at the beginning of the unit, so that students can begin to think about test items as they work through the unit. You might assign the assessment at least one week prior to the end of the unit and collect it two weeks after the unit is completed. If it is used as a group assessment, you may want periodically to schedule time for students to work with their groups on their tests.

Interdisciplinary Connection You may want to arrange for the science teacher in your school or some other knowledgeable person to speak to the class about the problems agriculturists have had in trying to control parasites with pesticides. This person might address the alternatives available.

Activity

In this experiment, you will investigate the answers to the following questions:

- How fast do bean sprouts grow?
- What happens to the growth of bean sprouts if the beans are placed in different solutions?

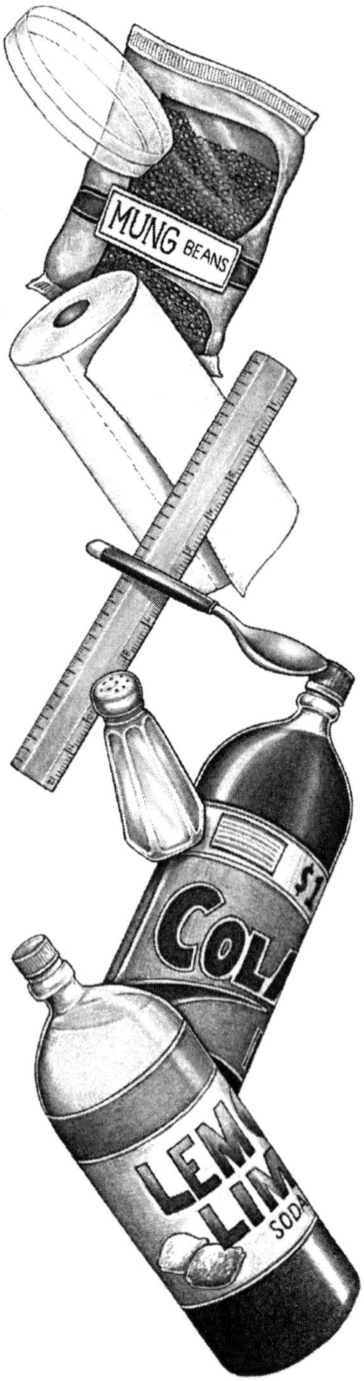

1. Before conducting an experiment, researchers hypothesize about, or predict, what will happen in the experiment. Read the description below and then make a prediction about the outcome.

Your group will need the following items:

- one petri dish
- paper towels
- 10 mung beans that have been soaked overnight in tap water
- a metric ruler for measuring in millimeters

You will also need one of the following solutions:

- tap water
- 1 teaspoon of salt per pint of water
- 1 fluid ounce of cola per pint of water
- 1 fluid ounce of lemon-lime soda per pint of water

Solutions and Samples
of student work

1. Answers will vary. Sample student responses:

   Bean sprouts grow best in tap water.

   Bean sprouts in salt water will not grow.

   Bean sprouts in different solutions will grow differently.

Hints and Comments

Materials cola (one fluid ounce per class); lemon-lime soda (one fluid ounce per class); metric rulers with millimeter markings (one per group of students); mung beans (10 per group of students); paper towels (three sheets per group of students); petri dishes (one per class); quart jars (four per class); salt (one teaspoon per class); scissors (one pair per group of students); tap water (one pint per class); teaspoons (one per group of students); thread or string (one segment per group of students)

Overview Students collect and prepare the materials needed for the experiment. They formulate hypotheses about the experiment.

Planning Students may work in small groups to set up the experiment. Provide a different solution for each group of students. You can use quart-size canning jars to hold the solutions. One jar of each solution is more than adequate for a class for the duration of the experiment. Each jar should be labeled with the name and concentration of the solution. You might want to place the jar for each solution next to the seeds growing in that solution.

Student Book page 2 explains how to prepare the four solutions. A solution of 2 teaspoons of salt per pint of water is a saturated solution. No bean sprouts grew in this solution in practice trials.

You might discuss with the class the reason why one solution is plain tap water. In a scientific experiment, a control is used. The growth of the mung beans in experimental solutions is compared to the growth of mung beans in plain water.

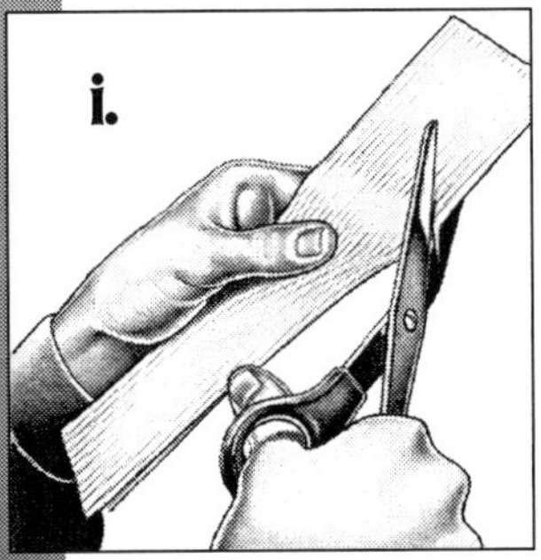
i.

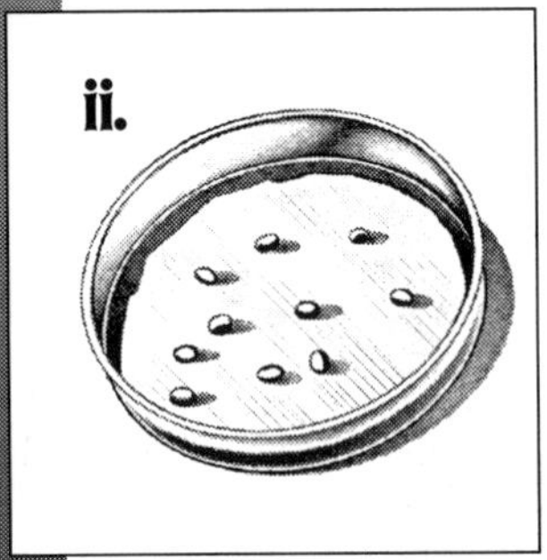
ii.

iii.

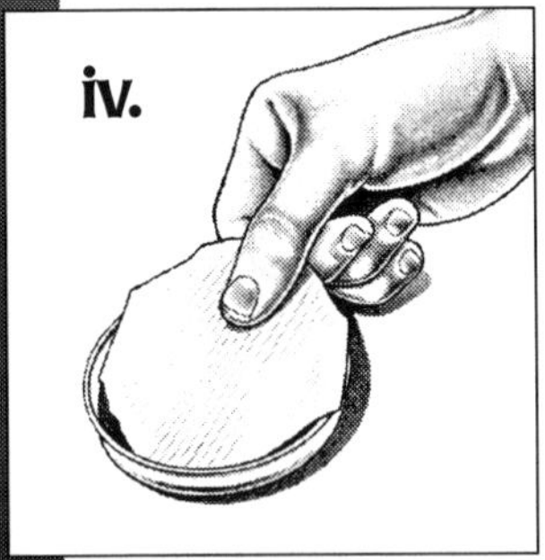
iv.

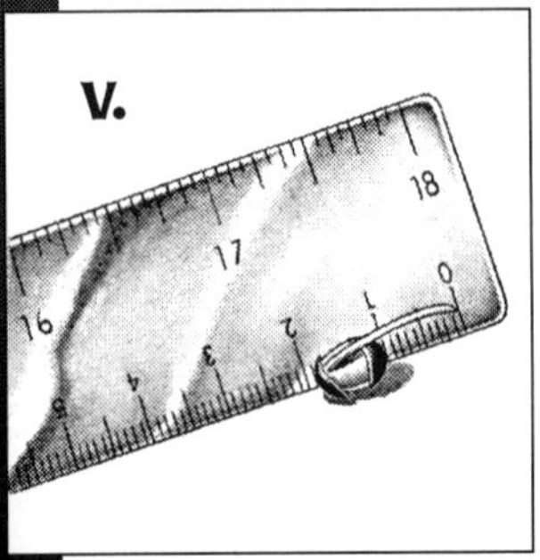
v.

Directions:

i. Cut three paper towels to fit the bottom of the petri dish.

ii. Put two layers of paper towels in the petri dish and arrange the mung beans on top. Figure out a way to identify the 10 individual beans so that you will be able to collect data for each bean as it grows.

iii. Soak the paper towels and beans by adding several teaspoons of your solution.

iv. Place another paper towel over the beans and dampen it with your solution. Place the cover on the petri dish. Label the petri dish with the name of your group and the type of solution used.

v. At approximately the same time each day, measure the length of each sprout in millimeters. Use the table on **Student Activity Sheet 1** to record the lengths of the sprouts and your observations about the growth of the beans. Keep track of the progress of the bean sprouts for seven days. During this time, add more solution as needed to keep the paper towels wet. You will use the results of this experiment later in the unit.

Bean Number	Length of Bean Sprout (in millimeters)							Observations and/or Problems
	Day 1	Day 2	Day 3	Day 4	Day 5	Day 6	Day 7	
1								
2								
3								
4								
5								
6								
7								
8								
9								
10								

Materials Student Activity Sheet 1 (one per group)

Overview Students conduct the experiment and measure the lengths of the bean sprouts each day for one week. There are no problems on this page for students to solve.

Planning Daily measurements of the beans should be recorded for seven school days on Student Activity Sheet 1. The data from this experiment will be analyzed in Section D. Taking exact measurements is very important. Consistent measurements are needed to be able to draw conclusions across groups.

Bean sprouts curl when they grow, and students need to decide how to measure them. One suggestion is to gently hold string or thread close to the sprout, following its curls, and then measure the length of the thread with a ruler. Students also must decide what parts of the sprout to measure. For example, will they measure from the edge of the seed to the end of the leaves, include the length of the seed, or exclude the leaves? You might have a class discussion about this before students measure the sprouts. Emphasize the need to have consistent measurements if conclusions are to be drawn across groups. Since the sprouts are brittle, students must also decide whether to exclude a bean from the experiment if its sprout breaks.

The salt solution should be stirred or shaken before each use.

Bringing Math Home Students may conduct a similar activity at home, using either different concentrations of the solutions or different solutions altogether.

Living in CITIES

The United States Census Bureau regularly investigates what percent of people live in cities and what percent live in rural areas. The information also indicates the movement of people from cities to rural areas and vice versa.

2. Why do you think it is important to know if people are moving from cities to rural areas?

After the information is collected, the Census Bureau reports the percent of urban population by state and the per capita income for each state.

3. Explain the meaning of "per capita income for each state."

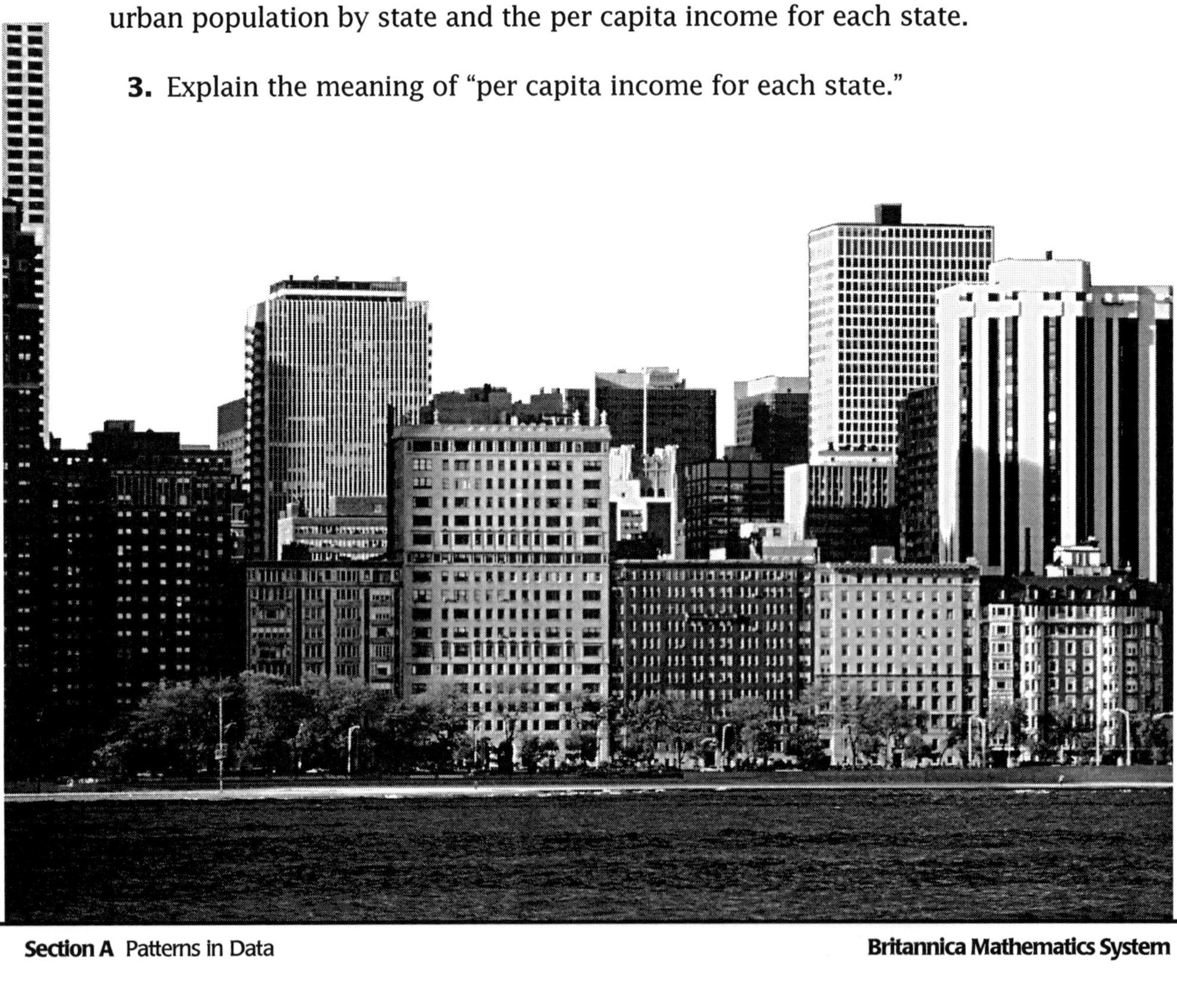

Solutions and Samples
of student work

2. Answers will vary. Sample responses:

A change in the number of people who live in cities will have an effect on the need for various city services.

New schools may have to be built in some areas.

There may be a greater demand for public transportation from rural areas to cities.

3. Answers will vary. Sample response:

Per capita income for a state is the average income for each person in that state. Per capita income is determined by dividing the total amount of income earned by all persons living in a certain area by the number of persons who live in the area.

Hints and Comments

Overview Students consider reasons why the U.S. Census Bureau gathers certain data. They also discover what per capita income is and how it is determined.

About the Mathematics In explaining the meaning of *per capita income,* students apply the concept of *mean* in a new context. The mean was introduced in the units *Picturing Numbers* and *Dealing with Data.*

Planning You may have students discuss the solutions to problems **2** and **3** as a class.

Comments about the Problems

3. The per capita income per state allows for a comparison between states. For example, $19,000 per capita income in Wisconsin compared to $20,500 per capita income in Minnesota means that the average person living in Wisconsin earned an income of $1,500 less per year than a person living in Minnesota.

You might want to make sure students know what is meant by the term *urban population.*

Extension You may have students investigate examples of other data that are gathered by the U.S. Census Bureau.

The *scatter plot* below and on **Student Activity Sheet 2** shows the results of the research conducted by the United States Census Bureau for the year 1992. The mean nationwide per capita income was about $20,000. Each state is identified as a data point in the plot labeled by its postal code. For example, Mississippi is represented by MS, and the District of Columbia is represented by DC.

Urban Population and Per Capita Income by State

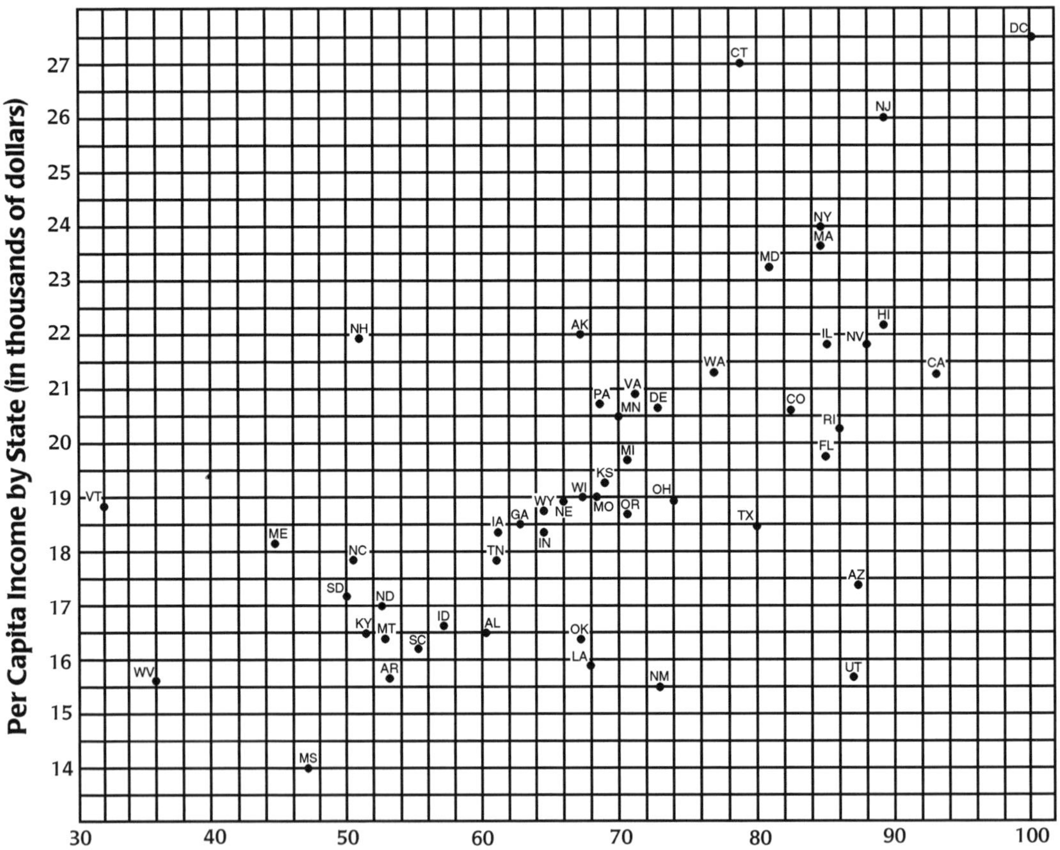

Source: Data from the U.S. Bureau of the Census, *Statistical Abstract of the United States: 1993* (113th edition.) Washington, DC, 1993

Use **Student Activity Sheet 2** to answer problems **4** and **5.**

4. Look at the data point for Utah (UT).

a. What percent of the people in Utah live in cities?

b. What other information is shown by this data point?

5. Find the data point for California (CA). Explain what that point represents.

Solutions and Samples
of student work

4. a. In Utah, 87% of the people live in cities.

b. The mean per capita income for Utah is about $15,700.

5. The data point for California indicates that about 93% of the people live in cities, and the per capita income of the people in the state is about $21,300.

Hints and Comments

Materials Student Activity Sheet 2 (one per student); transparency of Student Activity Sheet 2, optional (one per class); overhead projector, optional (one per class); maps of the United States, optional (one per student)

Overview Students read the scatter plot and explain what the data points for two states represent.

About the Mathematics Scatter plots are a way to graph data in order to investigate whether relationships between the two variables exist. Scatter plots are introduced in the unit *Dealing with Data.* If students are not familiar with scatter plots you may want to spend some time talking about this type of graph.

Planning Students may work on problems **4** and **5** individually or in small groups. *Note:* The District of Columbia is included in the data. There are 51 data points in the scatter plot.

Comments about the Problems

4. Encourage students to use a map of the United States to help them analyze the data in the scatter plot. They might consider, for example,

- the size of the cities in each state,
- the geographical features of each state, and
- the regions of the country.

5. Informal Assessment This problem assesses students' ability to analyze information from a scatter plot. If students are having difficulty locating California on the scatter plot, you might give them a hint that California has many urban areas.

Extension To help students gain experience reading scatter plots, you might ask them these additional questions:

Which place has the highest percent of urban population? [District of Columbia]

Which state has the lowest percent of urban population? [Vermont]

Which state has the lowest per capita income? [Mississippi]

Which place has the highest per capita income? [District of Columbia]

When interpreting the scatter plot of the Census Bureau data on page 5, Scott says:

6. In general, do you agree with Scott's statement?

Margaret studies the same scatter plot and makes the following comment:

7. a. Explain why so many states are below the mean. (*Note*: $20,000 is the correct mean.)

b. How can you describe the "typical" per capita income in another way?

6. Answers may vary. In general, this statement seems to fit the data in the graph.

7. **a.** Explanations will vary. Sample explanations:

    More people may live in the 18 states and the District of Columbia that are above the mean than in the 32 states that have per capita incomes below the mean.

    There may be some states that are outliers. These states have an unusually high per capita income, which is throwing off the mean.

   b. Answers will vary. Sample response:

    Another way to describe the typical per capita income is to use a median. Students can find the median from the plot by counting the data points from the top line of the scatter plot down until they reach the 26th state (26 is the median for 51 data points) or from the bottom line of the scatter plot up to the 26th state. They find Nebraska, at about $19,000. Half of the states have a per capita income greater than $19,000; half of the states have a per capita income less than $19,000.

Materials Student Activity Sheet 2 (one per student)

Overview Students comment on interpretations of the scatter plot. In doing so, they reflect on the meaning of the *mean.*

About the Mathematics Numerical measures to describe data—the mean, the mode, and the median—are introduced in the unit *Dealing with Data.*

Note: $20,000 is the nationwide per capita income (as described on page 5), not the mean of all 51 states' per capita income. One needs to be careful when finding the mean of means. In order to find the mean per capita income of all states, it is not correct to add up the per capita income of each state and divide by 51. This method does not take into account the differences in the population sizes of the states. See also problem **7a.**

Planning Students may work on problems **6** and **7** individually or in small groups. You may want to discuss problem **7a** as a class and then have students compare their answers to problem **7b.**

Comments about the Problems

6. In general, this statement seems to fit the data in the graph, although the relationship is very loose.

7. **a.** Populous states will have more influence on the mean per capita income for the whole United States than other states. For instance, the $21,400 of California has more influence than the $18,400 from Iowa, since the population of California is at least 10 times that of Iowa.

   b. The median income probably produces a more acceptable "typical" amount in this case. It might be helpful to rotate the page and count the number of states in each column. Asking students to find the median from the scatter plot can help them learn how to interpret data horizontally and vertically.

8. Dan, Eliza, and Yolanda are also studying the scatter plot of the Census Bureau data. Dan notices:

a. Consider Dan's statement. What else can you tell about New York as compared with Massachusetts?

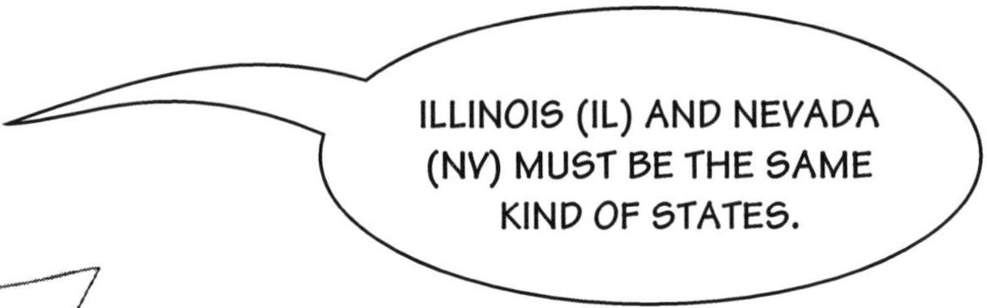

b. Comment on Eliza's statement. Do you agree?

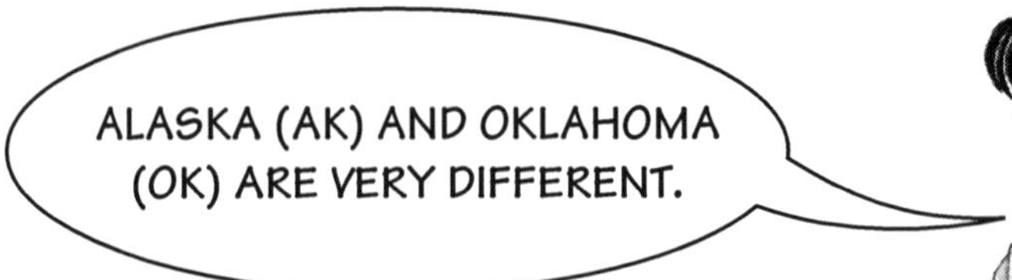

c. Do you agree with Yolanda? Explain your answer.

9. Locate the state in which you live on the scatter plot. Write a statement about the per capita income in your state and its relative position on the graph.

Solutions and Samples
of student work

8. Answers will vary. Sample responses:

a. New York has a higher per capita income than Massachusetts even though New York and Massachusetts have the same percent of urban population.

b. No. Illinois and Nevada have the same per capita income and the percents of urban population are about the same. However, it is inappropriate to conclude that the states are the same, since Illinois is one of the most populous states (Chicago is the third-largest city in the nation), and Nevada is one of the least populated states.

c. Yes. Alaska and Oklahoma have the same percent of urban population. However, Alaska has a significantly higher per capita income.

9. Answers will vary. Sample response:

About 74% of the population of Ohio lives in cities, and the per capita income is typical of states in the U.S., at about $19,000.

Hints and Comments

Materials Student Activity Sheet 2 (one per student)

Overview Students continue to study the U.S. Census Bureau data and investigate statements that compare states.

About the Mathematics In the statements on Student Book page 7, states are compared to each other. There are two variables that describe a state: mean per capita income and urban population.

When comparing items (or states) the reasoning behind a conclusion is more important than the conclusion itself. It is important for students to learn to justify and explain a statement or conclusion.

Planning Make sure students give clear explanations about the statements. Students may continue to work individually or in small groups on problems **8** and **9.** Problem **9** may be assigned as homework.

Comments about the Problems

9. Homework This problem may be assigned as homework. Students might look for other states that have the same per capita income or the same percent of urban population as the state in which they live. For example, students who live in Utah might compare Utah with New Mexico in terms of per capita income, and compare Utah with Arizona in terms of percent of urban population.

There seem to be two groups, or *clusters,* of states in the Census Bureau's scatter plot. The data point representing Georgia (GA) can be used to separate the points into two clusters.

10. Look at the cluster of data points to the left of Georgia.

a. What can you tell about the percent of people who live in cities and the per capita income in this cluster of states?

b. Are there any states to the left of Georgia that are *outliers* (states that lie outside the cluster)? Which states are they? What makes these states outliers?

11. Look at the cluster of data points to the right of Georgia. Identify the percent of people who live in cities and the per capita income in this cluster of states.

You could stop your investigation here. But sometimes you can find more information and make new statements by looking more closely at a graph.

© Kevin C. Rose

12. Review the scatter plot again. Write several statements that are more detailed than Scott's statement: "The higher the percent of people who live in cities, the higher the per capita income is for the state."

Solutions and Samples
of student work

10. a. Answers will vary. Sample response:

In the cluster to the left of Georgia, the per capita income ranges from $14,000 to $22,000. Most of the states in the cluster have a per capita income of between $16,000 and $18,000. The percent of people who live in cities is less than 62%.

b. Answers may vary. New Hampshire, Vermont, West Virginia, and Mississippi seem to be outliers.

In New Hampshire, about 51% of the population lives in cities, and the per capita income is $22,000 which is much higher than the typical per capita income for this cluster.

In Vermont, 32% of the population lives in cities, which is much less than the typical percent of urban population for this cluster, and the per capita income is close to $19,000.

In West Virginia, the percent of urban population is 36%, and the per capita income is about $15,500, which is slightly lower than typical in the cluster of states to the left of Georgia.

Mississippi is an outlier because of its very low income, the lowest of all the states.

11. Answers will vary. Sample response:

In the cluster to the right of Georgia, the percent of the urban population is over 64%, and the per capita income is higher than in states to the left of Georgia. In this cluster, most of the states have a per capita income of between $18,500 and $22,000.

12. Answers will vary. Sample student responses:

The states in the cluster to the left of Georgia have urban populations of less than 62%, with an average per capita income of about $17,000.

For the cluster of states to the right of Georgia, the pattern is very spread out. There seems to be a pattern showing that the higher the percent of people who live in cities, the higher the per capita income.

Other students may look for patterns related to regions of the country. One possible pattern is that, in general, northeastern states have higher per capita incomes than states in other regions of the country.

Hints and Comments

Materials Student Activity Sheet 2 (one per student)

Overview Students continue the exploration of the scatter plot and look for clusters of data points, outliers, and ways to describe the relationship between urban populations and per capita income.

About the Mathematics The term *outlier* is mentioned in the text. An outlier is a data point that is far from most of the values in the sample.

When looking for patterns in scatter plots, one can consider the following:

- If there are clusters (or groups) of data points, the population can be split into subgroups.
- If the points are all scattered, there is no relationship among the data points.
- If the points form a pattern such as a straight or a curved line, there is, to some degree, a relationship among variables. Investigations about the strength of correlations that appear to exist in the data are discussed later in this unit.

Planning You may have students work in small groups on problems **10** and **11.** Problem **12** may be used as an informal assessment.

Comments about the Problems

12. Informal Assessment This problem assesses students' ability to describe data numerically and to analyze representations of data. It also assesses their ability to draw conclusions based on given data sets and representations of data and to become aware of the questions that should be asked when analyzing data sets and representations of data.

It is important that students look closely at the graph to find more complex relationships. Critical review of data representations is required throughout this unit.

Writing Opportunity For problem **12,** you may have students write their detailed statements in their journals.

The United States Census Bureau categorizes states according to their geographical location. The map on this page shows these categories.

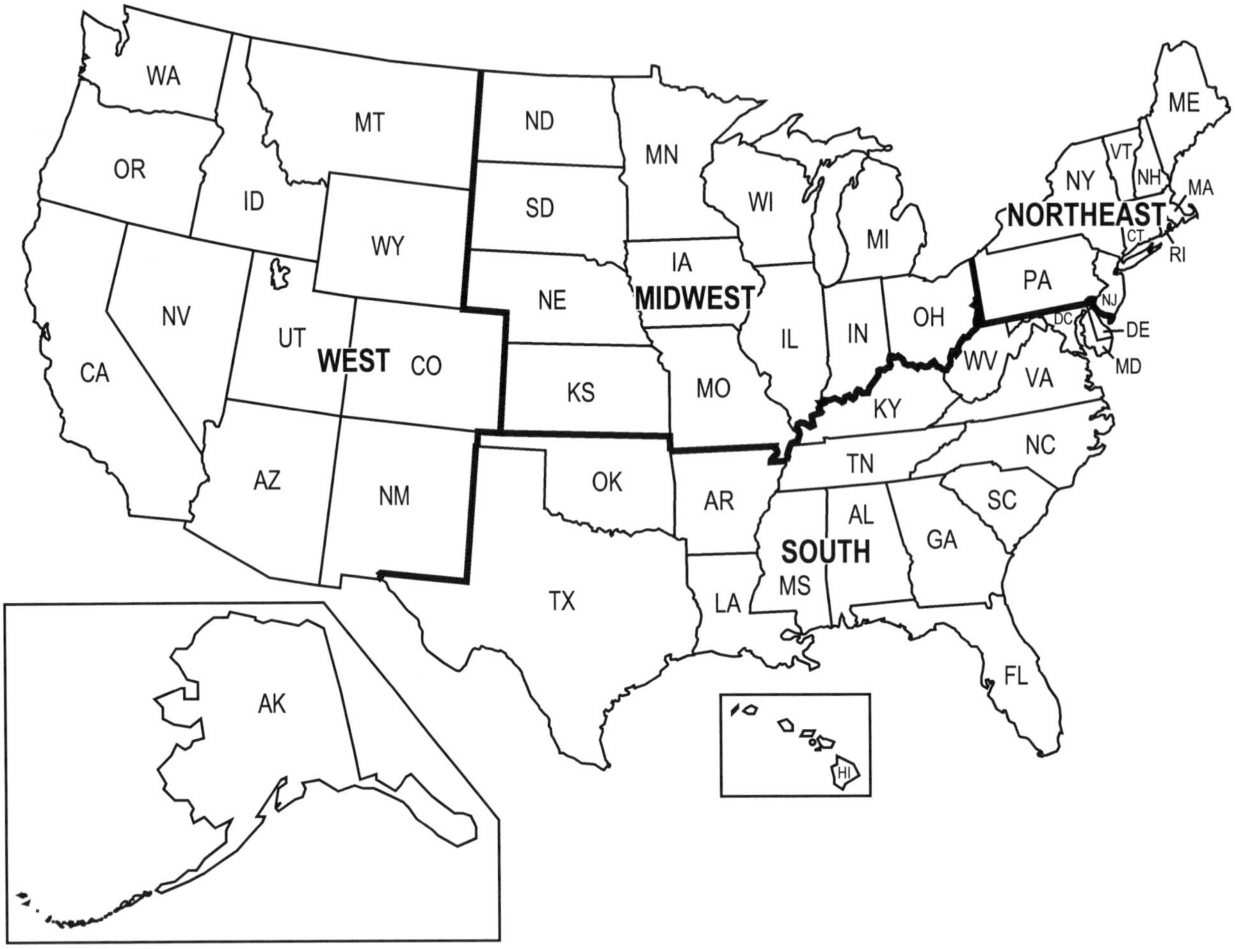

13. On the scatter plot on **Student Activity Sheet 2,** circle the dot for each state in the Midwest in blue, and circle the dot for each state in the South in red.

14. Explain the positions of Washington, D.C. (DC) and Maryland (MD) on the scatter plot. Compare their positions to those of other southern states.

15. Explain the position of Illinois (IL) on the scatter plot. Compare its position to those of other states in the Midwest.

13. See the scatter plot below. Instead of colors, different symbols are used.

Urban Population and Per Capita Income by State

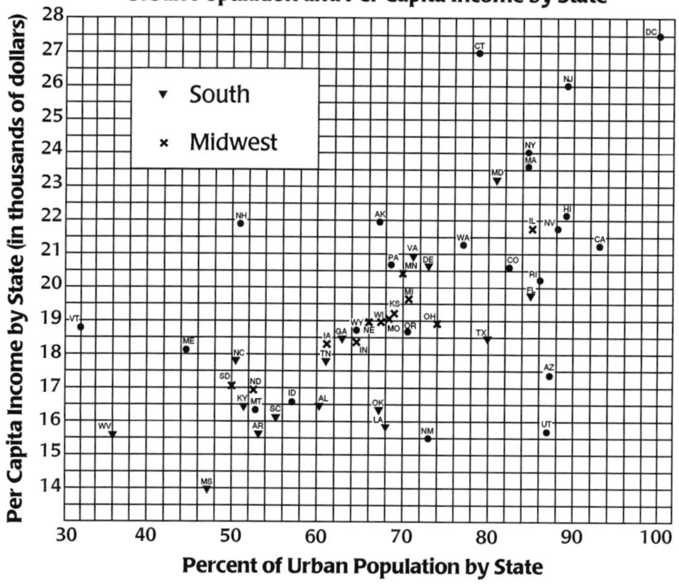

14. Explanations may vary. Sample response:

There are no rural areas in the District of Columbia. One hundred percent of the population lives in the city. Also, the per capita income is higher than in any other state. Because it is the nation's capital city, many professional people earn their living there, which drives up the area's per capita income.

Maryland is tiny, but it has a large city, Baltimore, and because it is so near the nation's capital, many people work in government and earn a good living.

Maryland and the District of Columbia are well removed from other Southern states on the scatter plot.

15. Explanations may vary. Sample response:

Illinois is well removed from other midwestern states on the scatter plot because its per capita income places it well above neighboring states. No doubt, this is due to Chicago, the third-largest city in the country.

Hints and Comments

Materials Student Activity Sheet 2 (one per student); blue and red markers (one of each per student)

Overview Students indicate states in the South and in the Midwest on the scatter plot and explain the positions of representative states from each area.

About the Mathematics On the previous page, the data were grouped in two clusters based on the picture in the scatter plot. Now the states are grouped according to geographic region. It appears that the data points for the Southern states are located along a straight line, describing a relationship. How to describe this linear relationship is examined in Section E.

Planning You may want to have students continue working on these problems in small groups.

Comments about the Problems

15. Informal Assessment This problem assesses students' abilities to analyze representations of data and to draw conclusions based on given data sets and representations of data. It also assesses their ability to become aware of the questions that should be asked when analyzing data sets and representations of data.

Summary

Data can be represented in a graph, such as a scatter plot like the one shown below. The data point *A* represents a car that weighs 1,960 pounds and gets 51 miles per gallon.

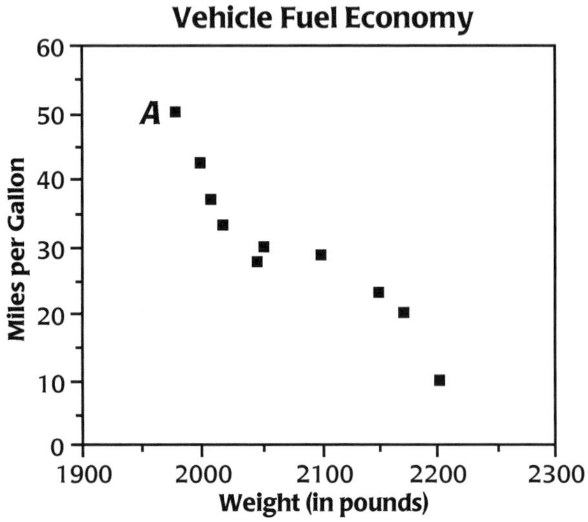

Some conclusions you draw from a graph may be very obvious. For example, a graph can show if there are clusters of data or outliers.

Other conclusions may require more complex explanations, such as a description of a typical data point.

Often, careful examination of a graph can raise new questions. More data gathering and research may be necessary to answer these new questions.

Summary Questions

16. Describe some features you might look for in a scatter plot. Why might these be important?

Solutions and Samples
of student work

16. Answers will vary. Sample student responses:

Look for clusters in the data. There may be something special about the data points in the clusters. For example, there may be a common underlying feature, such as geographic location.

Look for patterns. Does the scatter plot show a trend of some kind? Is there more than one pattern? If the data are clustered, does each cluster have its own pattern?

Look for outliers. There are characteristics of a state that make its data point an outlier. The District of Columbia, for example, is all urban.

Hints and Comments

Overview Students read the Summary and describe some of the characteristics and features of a scatter plot.

About the Mathematics A scatter plot (as introduced in *Dealing with Data*) is a helpful visual tool to describe data that come in pairs (two variables). Patterns in the scatter plot may suggest relationships. Sometimes, however, a third (or fourth) variable or factor that is not represented in the scatter plot is involved, and this extra variable may describe the pattern much better.

Students should develop a critical attitude toward representing data and interpreting represented data. They should always ask themselves questions about other possible factors or variables that may be involved, and they should ask themselves questions about how the data were collected. This issue is addressed in more detail in the next section.

Planning Students may work on problem **16** individually. You may use this problem for informal assessment. After students complete Section A, you may assign appropriate activities from the Try This! section, located on pages 58–62 of the *Insights into Data* Student Book, for homework.

Comments about the Problems

16. Informal Assessment This problem assesses students' ability to analyze representations of data.

Work Students Do

In this section, students discuss the process of sampling. They learn how to collect a random sample of the whole population and discover how the size of the sample might influence the outcome. Students interpret graphical representations of data and explore random numbers using a random-number table or a graphing calculator. They use a random-number simulation to investigate whether a stated percent for a given situation is reasonably accurate. Students will also investigate possible bias in survey results.

Goals

Students will:

- represent data graphically;
- describe data numerically;
- design, conduct, and analyze ways of data gathering: surveys, simulations, experiments;
- use random samples in gathering data;
- draw conclusions based on given data sets and representations of data;
- recognize possible bias in sample surveys;
- determine whether representations of data (numerical and visual) are appropriate;
- become aware of the questions that should be asked when analyzing data sets and representations of data.

Pacing

- approximately three 45-minute class sessions

Vocabulary

- bias
- probability
- random number
- random sample
- sample
- simulation
- trial

About the Mathematics

In a survey, a sample is selected from the population as a whole. Inferences about the population are based on the analysis of the data from the sample. In a random sample, each person has the same chance of being selected for the survey. Bias can result if the selection procedure for determining the sample was not objective and impartial. When the selection procedure is biased, taking a larger sample does not help. Many sources of bias are investigated in this section.

Also in this section, students use simulations to gather data. In order to examine the idea of how representative of a population random samples can be, students perform this experiment: they represent groups of 20 students in a state in which 80 percent of the people live in urban areas. The students are represented with random numbers from zero to nine. Before randomly selecting 20 numbers, however, students designate 8 of the 10 numbers as representing students from urban areas and 2 numbers as representing students from rural areas. Students learn that the more samples they add to the study, the more they begin to approach the 80 percent figure for urban dwellers.

Materials

- Student Activity Sheets 3 and 4, pages 145 and 146 of the Teacher Guide (one of each per student)
- Student Activity Sheets 10–12, pages 152–154 of the Teacher Guide, optional (one of each per student)
- graphing calculators, pages 33, 35, and 37 of the Teacher Guide, optional (one per student)
- overhead transparency of Student Activity Sheet 3, page 35 of the Teacher Guide, optional (one per class)
- overhead projector, page 35 of the Teacher Guide, optional (one per class)
- graph paper, page 37 of the Teacher Guide (two sheets per student)

Planning Instruction

You may want to start this section with a brief class discussion about how to collect good data. The questions posed on page 28 of the Teacher Guide can be used to direct the discussion.

Students may work on problems 1–3 and problem 10 in small groups. Students may work on the remaining problems individually.

Problems 11–16 are optional. If time is a concern, you may omit these problems or assign them as homework.

Homework

Problems 7 and 8 (page 36 of the Teacher Guide) may be assigned as homework. The Extensions (pages 29 and 41 of the Teacher Guide) may also be assigned as homework. After students complete Section B, you may assign appropriate activities from the Try This! section, located on pages 58–62 of the *Insights into Data* Student Book. The Try This! activities reinforce the key mathematical concepts introduced in this section.

Planning Assessment

- Problem 8 can be used to informally assess students' ability to represent data graphically and to design, conduct, and analyze ways of gathering data: surveys, simulations, experiments. This problem also assesses their ability to draw conclusions based on given data sets and representations of data and to become aware of the questions that should be asked when analyzing data sets and representations of data.
- Problem 10 can be used to informally assess students' ability to use random samples in gathering data and also to recognize possible bias in sample surveys. This problem also assesses their ability to determine whether representations of data (numerical and visual) are appropriate and to become aware of the questions that should be asked when analyzing data sets and representations of data.
- Problems 14 and 16 can be used to informally assess students' ability to represent data graphically, to describe data numerically, and to draw conclusions based on given data sets and representations of data.
- Problem 18 can be used to informally assess students' ability to recognize possible bias in sample surveys and to become aware of the questions that should be asked when analyzing data sets and representations of data.
- Problem 19 can be used to informally assess students' ability to design, conduct, and analyze ways of data gathering: surveys, simulations, experiments. This problem also assesses their ability to recognize possible bias in sample surveys and to become aware of the questions that should be asked when analyzing data sets and representations of data.

B. SELECTING SAMPLES

Collecting "Fair" Data

Data can be obtained from organizations such as the United States Census Bureau, and the results can then be graphed. But it is not always easy to get accurate data, as you may have seen in the units *Dealing with Data* and *Statistics and the Environment.*

Questions such as the following are important in statistics:

- How do you get reliable data?
- What is the best way to visually present the data?
- How do you draw accurate conclusions based on the data?

Looking carefully at graphical representations of data is important. Even graphs based on complete data, such as the scatter plot in Section A, must be studied carefully before reliable conclusions can be made.

Overview Students read questions about data collection, data representation, and drawing conclusions. There are no problems on this page for students to solve.

About the Mathematics As described in the three questions on this page, a statistical investigation can be split in three parts: data collection, data representation, and drawing conclusions. In this section, the focus is on how to get "good" samples. Determining what is good depends on the question to be answered. Often, a "good" sample is a representative sample of a larger group (the population). Representative in this context means that the sample has the same characteristics as the population, and that you can generalize for the whole population based on data collected for the sample.

Planning For an idea of how to begin this section, see the Planning Instruction section on page 27 of the Teacher Guide.

Extension You may ask students to look carefully at media such as TV, newspapers, and magazines for ways in which data are collected and represented. Students' observations should include whether or not the data collection resulted in good data, how the data were represented, and what conclusions could be drawn from the data.

The percent of people who live in cities and the per capita income by state are found by the Census Bureau.

In Texas, about 80% of the population lives in cities, as you can see from looking at the scatter plot on page 5 and on **Student Activity Sheet 2.**

Imagine you are visiting a band camp with students from all 50 states and Washington, D.C. There are 20 students from Texas at the camp.

Young musicians from around the world come to Interlochen Arts Camp in northern Michigan to study each summer. Photography by David Speckman.

1. Do you think there is a high *probability* that 16 of the 20 students from Texas live in cities?

Suppose you found that, in actuality, only eight out of the 20 Texas students live in cities.

2. Would this number surprise you? What are some possible reasons for the rather low number?

Solutions and Samples
of student work

1. Answers will vary. Sample response:

   Yes. In Texas about 80% of the population lives in cities. So, one would expect that 80% of the students selected from Texas live in cities. Eighty percent of 20 is 16.

2. Answers will vary, but some students may find this somewhat surprising. Some possible reasons may include the following:

   - The 20 students may not be typical of the students in the state. Many students may have come from one school in a rural area.
   - Twenty students is a small group to represent the population distribution of the state of Texas.

Hints and Comments

Overview Students investigate a context involving a group of 20 Texas band students, and they explore how representative this group is for a whole state.

About the Mathematics A critical problem in statistics is finding a way to specify a range of likely values for some event. In a situation where 80% of the population has some characteristic, one may expect that in a sample of 20, there would be about 16 persons having this characteristic. "About 16" should be interpreted as a number between 14 and 18; other numbers are possible, but less likely. The expectation of 16 assumes that the sample of 20 is representative of the greater population. However, this assumption needs to be checked carefully before accurate statements can be made about expectations. In the unit *Great Expectations,* students will more formally learn about expectations.

At this point in their study of statistics, students should begin to appreciate the need to think more deeply about problems in which statistics are used, developing a critical attitude toward the use of statistics.

Planning After a brief introduction in which the context is explained, you may have students work on problems **1–2** in small groups.

Comments about the Problems

1. Students may recognize that to have a representative sample that is exactly 80% of 20, or 16 people, is not likely, especially when they are dealing with such a small sample.
2. Students should reason that one or two, or even eight, students are also unlikely results. If the 20 students are typical of the rest of the students in Texas, students might think that 12 or 13 is a more reasonable answer.

Although you know that 80% of the Texas population lives in cities, it is difficult to predict how many live in cities for any group of 20 students. The *sample* is very small, and you have no additional information about the students in the sample.

If you randomly selected 20 students from Texas, the probability that only eight of them live in cities would be rather low. The probability that only one out of 20 lives in a city would be even lower.

One way to investigate each of the possibilities in more detail is to create a model, or *simulation*, of the situation.

The only tool you need is a set of *random numbers*. The set you will use consists of numbers from 0 to 9 in random order. You can read random numbers from a table or generate them on a calculator.

3. a. What do you think is meant by a "random number"?

b. Do you think the set of numbers below is a set of random numbers? Explain why or why not.

7, 7, 3, 0, 8, 4, 3, 0, 0, 1, 2, 7, 0, 7, 4, 1, 0, 2, 4, 2, 3

Solutions and Samples
of student work

3. **a.** Answers will vary. Students may think a random number is one that a person just happens to select or that occurs by accident.

   b. Answers will vary. This sequence of numbers may or may not be a set of random numbers. The absence of 5s, 6s, and 9s may suggest that it is not. However, there is no guarantee that such a small set will represent every possible number; some numbers may appear more often than is expected.

Hints and Comments

Overview Students continue to investigate probability in the context of the 20 Texas band students. They also are introduced to random numbers.

About the Mathematics In the unit *Dealing with Data,* the concepts of *sample, population,* and *representative sample* are introduced. Students are expected to be somewhat familiar with these terms.

On this page, the term *random number* is introduced. A random number is a number that is chosen with no preference, thus with as much of a chance of being chosen as any other number.

Planning Students may continue to work in small groups on problem **3.**

Comments about the Problems

3. **a.** Allow students to describe *random* in their own words rather than providing them with a formal definition.

   b. It is difficult to tell whether a given set of numbers is random, especially when it is a small set of numbers. One way to see whether a set of numbers is random is to make a histogram and observe whether the frequencies for each digit balance out as the set of random numbers increases. Each digit will occur approximately the same number of times. There is no rule for how large this set should be.

You can use random numbers to simulate the number of students who live in cities using the following method:

- You know that about 80% of the population of Texas lives in cities.
- Let 1, 2, 3, 4, 5, 6, 7, and 8 each represent a student who lives in a city. (These eight digits will represent 80%.)
- Let 9 and 0 each represent a student who does not live in a city. (These two digits will represent 20%.)

To use a random number table, such as the one on **Student Activity Sheet 3,** begin by placing it in front of you. Close your eyes and point your pencil at the random number table. Use the first digit you point to with your pencil as your first number. Write this number down. Then decide whether you are going to move to the right, left, up, or down on the table. Once you have decided, move in that direction and write down the next 19 digits in the order in which you choose them. If you run out of numbers in the row or column, move to the next row or column and reverse your direction.

4. Use the random number table on **Student Activity Sheet 3,** or use your calculator (See **Student Activity Sheet 10**), to select 20 random digits. Each of these 20 digits will represent a student in the sample. For every 1, 2, 3, 4, 5, 6, 7, or 8 you select, record a Y for yes. For every 9 or 0 you select, record an N for no.

An example of how to tally the number of yes's and no's is shown in the table on the left.

	Number of Yes's	Number of No's
Sue	16	4
Tad	10	10
Tony	14	6
Kua	13	7

5. a. Make a table like the one on the left to tally the number of yes and no results for the entire class.

b. What does the number of yes's you recorded represent?

c. How do your results compare with your classmates' results? Explain any similarities or differences.

4. Results will vary. Sample results:
   - 9, 6, 6, 0, 8, 5, 7, 4, 2, 0, 2, 7, 5, 1, 4, 0, 3, 3, 6, 0
     15 yes's, 5 no's
   - 3, 5, 9, 1, 3, 4, 2, 0, 7, 5, 6, 3, 0, 9, 6, 0, 4, 5, 9, 9
     13 yes's, 7 no's

5. **a.** Tables will vary, depending on students' data.

   b. The number of yes's represents the number of students who live in the cities.

   c. Answers will vary, depending on students' data.

Materials Student Activity Sheet 3 (one per student); Student Activity Sheet 10, optional (one per student); graphing calculators, optional (one per student); overhead transparency of Student Activity Sheet 3, optional (one per class); overhead projector, optional (one per class)

Overview Students use a graphing calculator or a random-number table to find 20 random numbers to represent a group of 20 students.

About the Mathematics Designing the simulation is an important activity. You need to think about what the experiment should look like and define what each of the outcomes means. After the design, the experiment can be carried out.

In problem **5,** it will probably become clear that the more data collected, the more the distribution will look like the expected distribution.

Planning You may want to begin with a class discussion about how to find random numbers and how to set up an experiment to simulate a situation. *Note:* Student Activity Sheet 10 has directions for finding random numbers with a graphing calculator. Students may work individually on problems **4–5.**

Comments about the Problems

4. Using a random-number table may not be intuitive. Complete a few examples with the class. An effective method is to make a transparency of the random-number table and ask a student to come to the overhead projector to complete the process of determining a set of random numbers. All students can tally the yes and no results and record the results in the table.

Students can choose their movements in the table. A student might think, "After I put my pencil on a digit, I will move down a column to get my numbers. If my column ends, I will go to the left one column and then go up that column until I get 20 digits."

Number of Yes's	Number of Students
10	
11	
12	
13	
14	///
15	
16	
17	

If there are 25 students in your class, you have taken 25 sample trials. In other words, each one of you has looked at one sample of 20 band students to determine how many live in cities.

There is another way to organize the class results. In the table on the left, the number of yes's, and the number of students who had that number of yes's, are recorded.

6. a. What do the three tally marks in the table mean?

b. Use the data collected in problem **5** to make a new table like the one on the left.

7. a. On **Student Activity Sheet 4,** use the data from the table you made in problem **6b** to make a histogram.

b. Use the data to determine the number of yes's that are likely in 25 trials.

c. In 25 trials, how likely do you think it is to have 14 to 18 yes's?

d. What do the results of your simulation tell you about the number of Texas students at the band camp who are likely to live in cities?

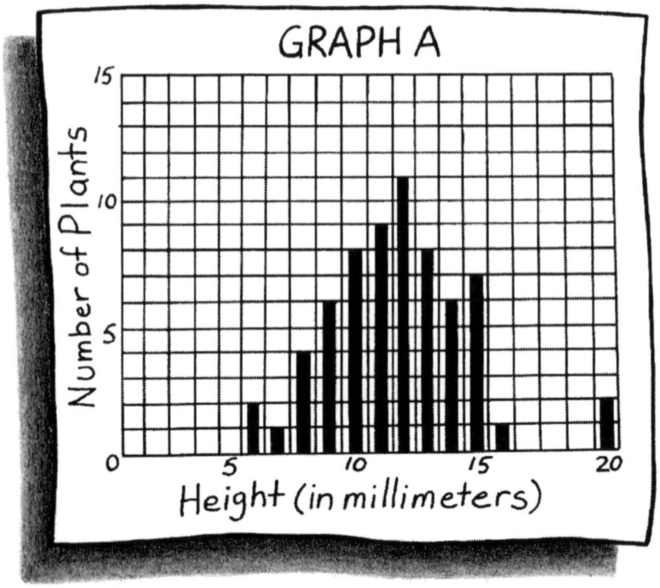

8. Repeat the simulation you did in problem **4.** Collect all data from the class. Add the new results to the table you made in problem **6b.**

a. Make a histogram using the new table. How does this histogram compare with the first one?

b. What kind of results do you think you would get if you continued to repeat the experiment?

Solutions and Samples
of student work

6. a. The three tally marks indicate that three students had exactly 14 yes's.

b. Tables will vary. Sample table:

Number of Yes's	Number of Students
10	
11	//
12	/
13	///
14	//
15	卌 /
16	///
17	卌 //

7. a. Histograms will vary. Sample histogram:

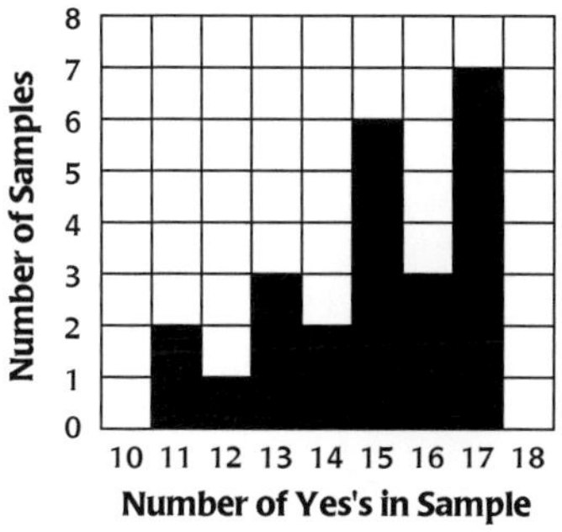

b. Answers will vary. In the sample above, the mean is 15.

c. Answers will vary. Obtaining 14 to 18 yes's is very likely but the number from one trial to another will vary.

d. Answers will vary. The results of the simulation most likely will indicate that from 14 to 18 of the students might live in cities.

8. a. Graphs will vary, depending on the data accumulated by the class. Sample histogram:

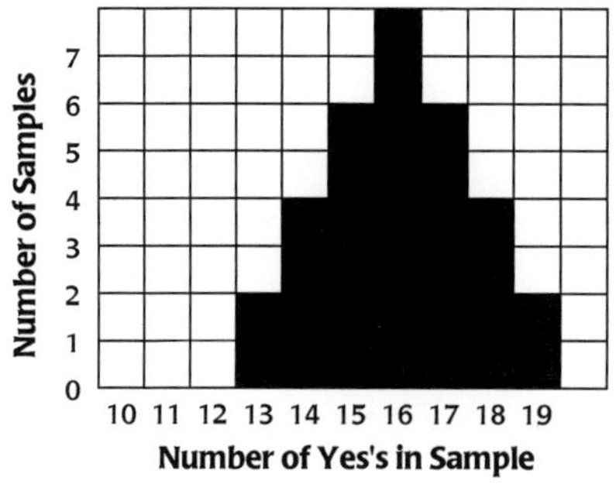

b. The more often the experiment is repeated, the more likely that the result is 16 yes's. See the histogram above.

Hints and Comments

Materials Student Activity Sheet 4 (one per student); graph paper (two sheets per student); graphing calculators, optional (one per student); Student Activity Sheet 11, optional (one per student)

Overview Students collect data from all students in the class, graph these data, and draw conclusions based on their graphs.

About the Mathematics As students work through the problems on this page, the effect of larger samples will become apparent.

Planning Students may work on problems **6–8** individually. You may use problem **8** as an informal assessment. Problems **7** and **8** may be assigned as homework. *Note:* Student Activity Sheet 11 gives directions for using a graphing calculator to make histograms.

Comments about the Problems

6. If you have fewer than 25 students in your class, you might complete the experiment a second time to obtain a larger sample. Too few trials may not give enough data for students to see the patterns emerging.

7–8. Homework These problems may be assigned as homework.

7. a. Student data can also be recorded in a number-line plot or a line graph (as introduced in the unit *Dealing with Data*).

8. Informal Assessment This problem assesses students' ability to represent data in a histogram and to design, conduct, and analyze ways of gathering data. This problem also assess their ability to draw conclusions based on given data sets and representations of data and to become aware of the questions that should be asked when analyzing data.

b. When the data are graphed for large numbers of students (when the experiment is done many times), a nearly bell-shaped distribution with the center at 16 will develop.

In the simulation you just completed, you found that *random samples* reflect the population. By taking a number of samples from a population with 80% yes's, you found that 14 to 18 yes's out of 20 was a likely result. In the particular sample introduced on page 12, only eight out of the 20 students lived in cities. You can conclude that this sample of 20 students was not typical of the population of Texas.

Many samples are *biased*, because they favor certain outcomes, or they favor some parts of the population over others. Care must be taken so that any member of the population has an equally likely chance of being chosen in the sample. Consider the following situation:

Some TV stations poll the public. Viewers are urged to call specific numbers to voice their opinions. Dialing one number registers a yes vote; dialing another number registers a no vote.

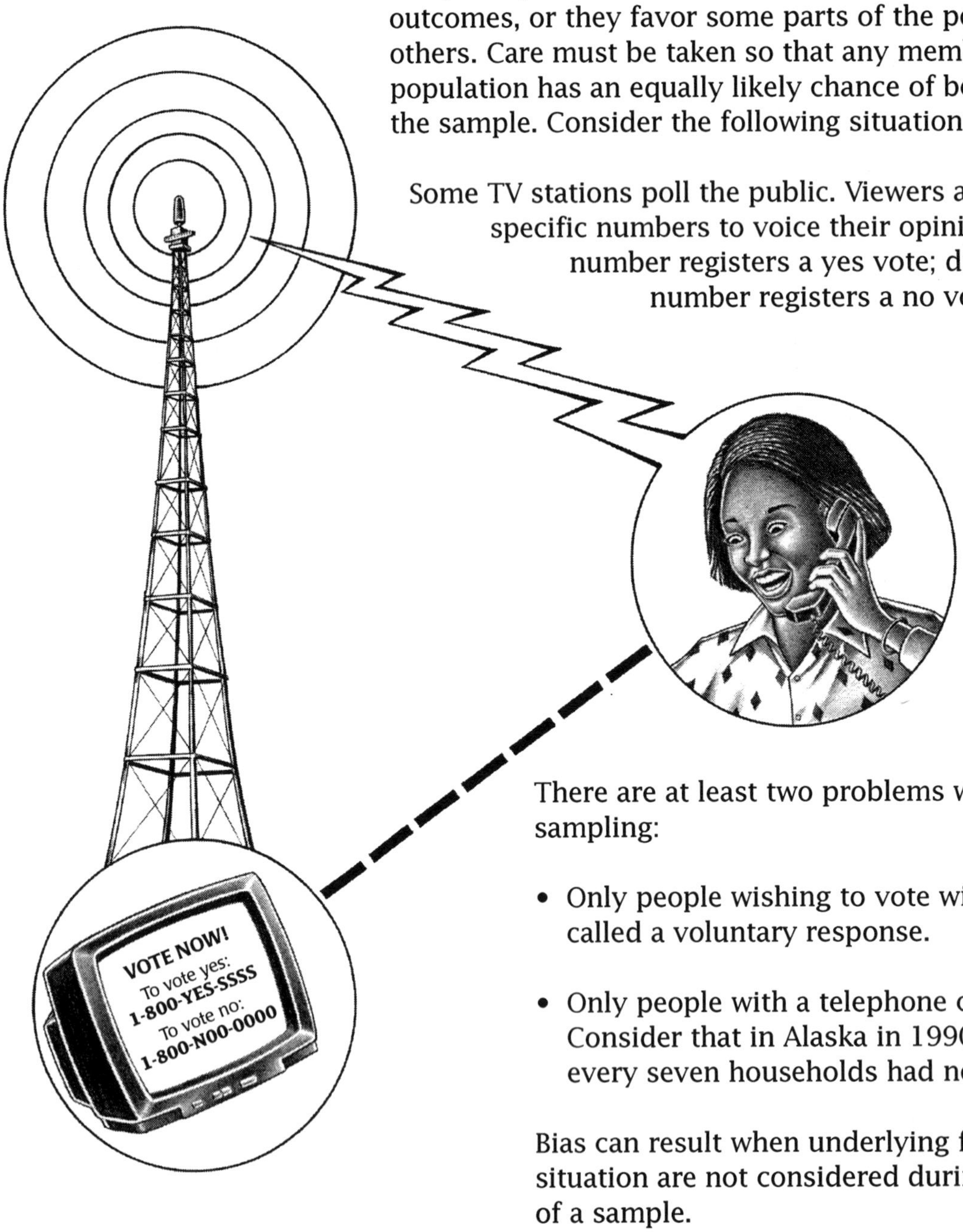

There are at least two problems with this type of sampling:

- Only people wishing to vote will call. This is called a voluntary response.
- Only people with a telephone can respond. Consider that in Alaska in 1990, one out of every seven households had no phone.

Bias can result when underlying factors about a situation are not considered during the selection of a sample.

9. Why might the information collected about students at a band camp be biased?

Solutions and Samples

of student work

9. Answers will vary. Sample responses:

- The students may be from the same school.
- The sample of students may not be a random sample.
- Students often go to camps with their friends, so they would live in the same place.
- Twenty is a small sample size.

Hints and Comments

Overview Bias and underlying factors in sampling techniques are discussed.

About the Mathematics Bias in a sample is a common phenomenon in statistics. It is almost impossible to have a completely unbiased sample, but one should consider very carefully what factors might cause a relevant bias before collecting data.

Planning You may want to use the text as the basis for a discussion of bias in sampling. Then, ask students to give their opinion of how bias may affect the results of surveys and polls conducted by newspapers and magazines. Students may work on problem **9** individually.

Comments about the Problems

9. Students should explain why the information about the student band members could be biased. You might want to collect students' answers and discuss which reasons cause more bias than others and which type of bias is more important than others.

10. Read each of the following survey situations carefully. Explain how each poll could involve bias:

a. The Chief of Police in a major U.S. city wants to determine how the public feels about the department. He prepares a questionnaire and sends police officers out to interview people in randomly selected sections of the city.

b. A magazine for health foods and organic healing wants to establish that large doses of vitamins improve health. The editor asks readers who have regularly taken vitamins in large doses to write to the magazine and describe their experiences. Of the 2,754 readers who reply, 93% report some benefit from taking large doses of vitamins.

10. Answers will vary. Sample responses:

a. Since the police officers are interviewing the public about police officers, people may feel uncomfortable relating their true feelings and may choose not to respond; this would be an example of bias based on a high nonresponse rate. Bias may also occur if people answer in ways they feel the officer wants, rather than sharing their true feelings.

b. This is an example of bias based on voluntary response. The people who responded took vitamins regularly, so it is likely that they consider their experiences with vitamins to be positive. Bias may also occur when people interpret words differently. For example, what does "improvement in health" mean? What is "a large dose" of vitamins? Also, the population reading the magazine may not be representative of the whole population.

Overview Students investigate two situations involving bias.

About the Mathematics There are different kinds of bias addressed in these problems:

- a high rate of nonresponse
- surveys not conducted by an appropriate person
- voluntary response
- asking the wrong questions, suggesting only positive (or negative) responses
- selecting a poor sample (not representative for the whole population)

Planning Problem **10** extends over two pages. Students may work in small groups. This problem may be used as informal assessment.

Comments about the Problems

10. Informal Assessment This problem assesses students' ability to use random samples in gathering data and to recognize possible bias in sample surveys. This problem also assesses their ability to determine whether representations of data (numerical and visual) are appropriate and to become aware of the questions that should be asked when analyzing data.

Extension On Teacher Guide page 29, an activity was suggested in which students investigate media resources for examples of data collection. You may remind students of this activity and ask them to investigate examples of data collection in which bias is involved.

10. c. A researcher wants to find out how many Americans intend to vacation in the United States one year. To avoid bias, she selects 27 travel agencies in large cities and interviews every seventh visitor. The results of her research are published and titled "Record Number of Americans to Foreign Destinations."

d. In 1936, the largest poll about the presidential election between Franklin Roosevelt and Alf Landon was taken by a magazine called *Literary Digest.* The publisher sent out 10 million questionnaires to people listed in telephone books. They also used other sources, such as car registrations and its own subscriber lists. The magazine received a world record 2.4 million replies. As a result of the poll, *Literary Digest* predicted that Landon would win by a margin of 57% to 43%. However, Roosevelt won the election. Another research group used a much smaller sample of 50,000 people and predicted correctly that Roosevelt would win the election.

Solutions and Samples
of student work

10. c. Answers will vary. Sample responses:

This is an example of bias resulting from using a poor sample. Americans who go on vacations in the United States may not use the services of a travel agent, while those traveling overseas might need such a service. Also, travel agencies in large cities might receive more requests for foreign travel than agencies in small cities.

d. This is an example of bias resulting from a sample that was selected by an inappropriate sampling technique. *Literary Digest* sampled people who had telephones or cars. In 1936, few Americans owned these items. *Literary Digest* also used its subscriber list to select poll participants. The subscribers may not be representative of the whole population. The other research group probably used a more representative sample of the population.

Hints and Comments

Overview Students investigate two more situations involving bias.

Did You Know? Sometimes a biased sample is chosen on purpose to manipulate the outcomes of a survey. In Darrell Huff's *How to Lie with Statistics* (New York: Norton & Company, 1954) several examples of the abuse of statistics are described. Huff also provides examples of poor sampling.

Suppose you looked at a random set of 50 numbers.

11. a. How many numbers in the set would you expect to end with a 5 or a 9? Why?

b. Would you expect to get exactly this many numbers ending with a 5 or a 9 every time you looked at a random set of 50 numbers?

12. Suppose the set of 50 numbers you looked at were the prices of items at a supermarket. How many would you now expect to end in a 5 or a 9?

13. Collect a sample set of 50 prices from a supermarket. Count the number of prices that end in a 5 or a 9. Compare this number with your answer to problem **12.**

Supermarket Prices

Number of Prices Ending in a 5 or a 9	Number of Students

14. a. Make a table like the one on the left to collect and record the class supermarket data.

b. Make a plot of the data in the table.

c. How does the data compare with your answers to problems **11b** and **12?**

According to *Harper's Index*, approximately 80% of supermarket prices end in a 5 or a 9. The number of prices ending in a 5 or a 9 will vary for different sets of 50 items.

15. a. Design a simulation using random numbers like you did on page 14 to represent prices so that 80% of them end in either a 5 or a 9. Explain how you designed it.

b. Use the random number simulation you designed to generate 50 numbers.

c. In your simulation, how many of the prices end in a 5 or a 9?

16. a. Make a new table by combining the data simulated by all students for problem **15.** Make a plot of the class results.

b. What information can you get from your new plot?

c. Compare your simulated results (problem **15b**) with the results from the entire class.

17. Compare the results of the simulation in problem **15** to the results of the actual samples you collected from the supermarket in problem **13.** Do you think that the 80% figure used by *Harper's Index* is correct? Explain your reasoning.

Solutions and Samples

of student work

11. a. 20%. Explanations will vary. Sample explanation:

The last digit could be any number between 0 and 9. Each one of these numbers would have a 10% chance of being the last digit, so there is a 20% chance that a 5 or a 9 is the last digit.

b. No, the number is likely to vary.

12. Answers will vary. Some students may know that many prices in the supermarket have a last digit that is a 5 or a 9 and would therefore expect more than 20% to end in a 5 or a 9.

13. Answers will vary, depending on student data.

14. a. Tables will vary depending on the data collected.

b. Answers will vary. Students may use a box plot or histogram.

c. Answers will vary depending on student data.

15. a. Answers may vary. Sample response using the simulation procedure developed in this section:

Let 1, 2, 3, 4, 5, 6, 7, and 8 represent items with prices ending in a 5 or a 9.

Let 9 and 0 represent items with prices that do not end in a 5 or a 9.

Record a yes whenever a 1, 2, 3, 4, 5, 6, 7, or 8 is selected.

Record a no whenever a 9 or 0 is selected.

b–c. Answers will vary depending on student data, but will be close to 80%.

16. a. Answers will vary depending on students' answers to problem **15.** The result should be that about 80% of the prices end with a 5 or 9.

b. Answers will vary. Sample response:

The larger the set of data, the closer the distribution is getting to 80% of the prices ending in a 5 or 9.

c. Answers will vary. Students may recognize that if their simulation does not fit the plot, it is probably "canceled out" by someone else's simulation.

17. Answers and explanations will vary. Students may say yes if the distributions of prices from the supermarket and from the simulations are about the same.

Hints and Comments

Materials Student Activity Sheet 12, optional (one per student)

Overview Students use their knowledge of simulating with random numbers in a new context.

Planning Students may work on problems **11–17** individually. Problems **14** and **16** may be used as informal assessment. The problems on this page are optional. *Note:* Student Activity Sheet 12 gives directions for using a graphing calculator to make box plots.

Comments about the Problems

12. a. Supermarket prices often end with the number 9 because a price such as \$2.99 seems less expensive than \$3.00.

13. Discuss ways in which students could select a diverse sample of 50 prices in a supermarket. One way is to choose a diverse sampling of products from different aisles.

14–16. Informal Assessment These problems assess students' abilities to represent data, describe data numerically, and to draw conclusions based on data.

15. In one sample of 50 numbers, 42 of the numbers represented items with prices ending in a 5 or a 9.

16. c. If students have an answer that is very different, check their simulation design.

17. It could be that the claim by Harper's was incorrect. More checking should be done before accepting their claim.

As a class, discuss the factors that may have affected the results of sampling the actual supermarket prices. For example, the sample prices collected by the class may not have been a diverse selection of the prices at the supermarket. Different supermarkets may have different prices.

Summary

A population is a group of people or set of objects you want to gather information about.

You can collect data by questioning a small sample of people from a specific population, or by examining a small sample of objects from a set that has like characteristics.

When taking a sample, it is important to do so randomly, so that every member of the population has an equal chance of being selected.

You can also collect data by designing and running an experiment, or simulation.

When conducting surveys, you should avoid bias. Some possible causes of bias are:

- incorrectly choosing the sample,
- neglecting to account for the people who do not respond,
- letting interviewers select the people they want to interview.

Summary Questions

18. Select one example from this section that you think best represents each of the causes of bias given above.

19. Some people think that the larger the sample you take, the less chance you have of bias. Do you agree? Explain your answer.

18. Answers will vary. Sample responses:

Incorrectly choosing the sample:

- telephone survey in Alaska
- survey of people who use travel agents in the United States
- political poll in 1936 of people who had telephones or cars

Neglecting to account for the people who do not respond:

- health food magazine surveying readers who regularly take vitamins

Letting interviewers select the people they want to interview:

- responses by people who take vitamins regularly
- people being interviewed by police officers

19. Answers will vary. Sample response:

If the sample is not taken properly to ensure representativeness, then bias will occur regardless of the size of the sample.

Overview Students read the Summary and reflect on possible causes of bias.

Planning You may have students work on problems **18** and **19** individually. These problems may be used as informal assessment. After students complete Section B, you may assign appropriate activities from the Try This! section, located on pages 58–62 of the *Insights into Data* Student Book, for homework.

Comments about the Problems

18–19. Informal Assessment These problems assess students' ability to recognize possible bias in sample surveys and to become aware of the questions that should be asked when analyzing data sets and representations of data.

Students should recognize the need for further investigations in biased situations. Two things they might consider are using a larger sample and investigating how the sample is chosen.

19. Informal Assessment This problem assesses students' ability to design, conduct, and analyze ways of data gathering: surveys, simulations, experiments.

Work Students Do

Students investigate graphs to determine whether they represent data clearly and accurately. First they encounter a graph with no quantitative information in it and, consequently, no scaling of the axes. Then students look at a pair of graphs that present the same data, but whose *y*-axes are scaled differently and begin with different numbers. Students draw a graph to represent data from a table on world population figures and then discuss the inconsistent scale used in a given line graph. Histograms and pictographs are then explored for misleading representations of data. And, finally, students design their own statistical surveys and represent the data graphically.

Goals

Students will:

- represent data graphically;
- describe the relationship between two variables;
- design, conduct, and analyze ways of gathering data: surveys, simulations, experiments;
- use random samples in gathering data;
- analyze representations of data;
- draw conclusions based on given data sets and representations of data;
- recognize possible bias in sample surveys;
- determine whether representations of data (numerical and visual) are appropriate;
- become aware of the questions that should be asked when analyzing data sets and representations of data.

Pacing

- approximately two to three 45-minute class sessions

Vocabulary

- origin
- scaling

About the Mathematics

This section introduces misleading graphical representations of data. Some common methods for misrepresentation are using inconsistent scaling of the axes, not beginning each axis at the origin, and using dimensions for pictographs or bar charts that do not match the data. Each of the methods can make a change in data values seem dramatic or minimal, depending on the desired effect. (See the charts on Student Book page 22 for an example of changing the effect of a graph with a change in scaling.)

Materials

- Student Activity Sheet 5, page 147 of the Teacher Guide (one per student)
- transparency of Student Activity Sheet 5, page 55 of the Teacher Guide, optional (one per class)
- overhead projector, page 55 of the Teacher Guide, optional (one per class)
- graph paper, pages 55 and 63 of the Teacher Guide (two sheets per student)
- graphs from a newspaper or magazine, page 65 of the Teacher Guide (one per student)
- blank paper, page 65 of the Teacher Guide (one sheet per student)
- glue or tape, page 65 of the Teacher Guide (one per student)

Planning Instruction

The primary goal of this section is for students to look more closely at data sets and representations of data. You may want to begin the section with a brief discussion about data and its representation. You might also make transparencies of various line graphs, bar graphs, charts, and tables and display them on an overhead projector. Encourage students to look closely at these representations. Ask questions such as, *Are the scales on the axes consistent and labeled appropriately? In a bar graph, what does each bar represent? In a line graph, what does the line represent? Are there any trends in the data? Does the graph give a misleading impression?*

Students can work on problems 1–5 and the activity on page 62 in small groups. Problem 6 may be done as a class. Students may work on the remaining problems individually.

Problems 8 and 9 are optional. If time is a concern, you may want to skip these problems or assign them as homework.

Homework

Problem 6 (page 56 of the Teacher Guide) can be assigned as homework. The Bringing Math Home activity (page 53 of the Teacher Guide), the Extensions (pages 57 and 65 of the Teacher Guide), and the Writing Opportunity (page 63 of the Teacher Guide) also can be assigned as homework. After students complete Section C, you may assign appropriate activities from the Try This! section, located on pages 58–62 of the *Insights into Data* Student Book. The Try This! activities reinforce the key mathematical concepts introduced in this section.

Planning Assessment

- Problems 5 and 13 can be used to informally assess students' ability to analyze representations of data and to determine whether representations of data (numerical and visual) are appropriate.
- Problem 7 can be used to informally assess students' ability to describe the relationship between two variables and to analyze representations of data. It also assesses their ability to recognize possible bias in sample surveys and to determine whether representations of data (numerical and visual) are appropriate.
- Problem 12 can be used to informally assess students' ability to design, conduct, and analyze ways of gathering data: surveys, simulations, experiments; to use random samples in gathering data; and to recognize possible bias in sample surveys.
- Problem 14 can be used to informally assess students' ability to represent data graphically and to design, conduct, and analyze ways of gathering data: surveys, simulations, experiments.

C. INTERPRETING GRAPHS

Different Impressions

Graphs are useful for representing information in a clear and concise way. Shown below is a good example of a graph that conveys information using only two words.

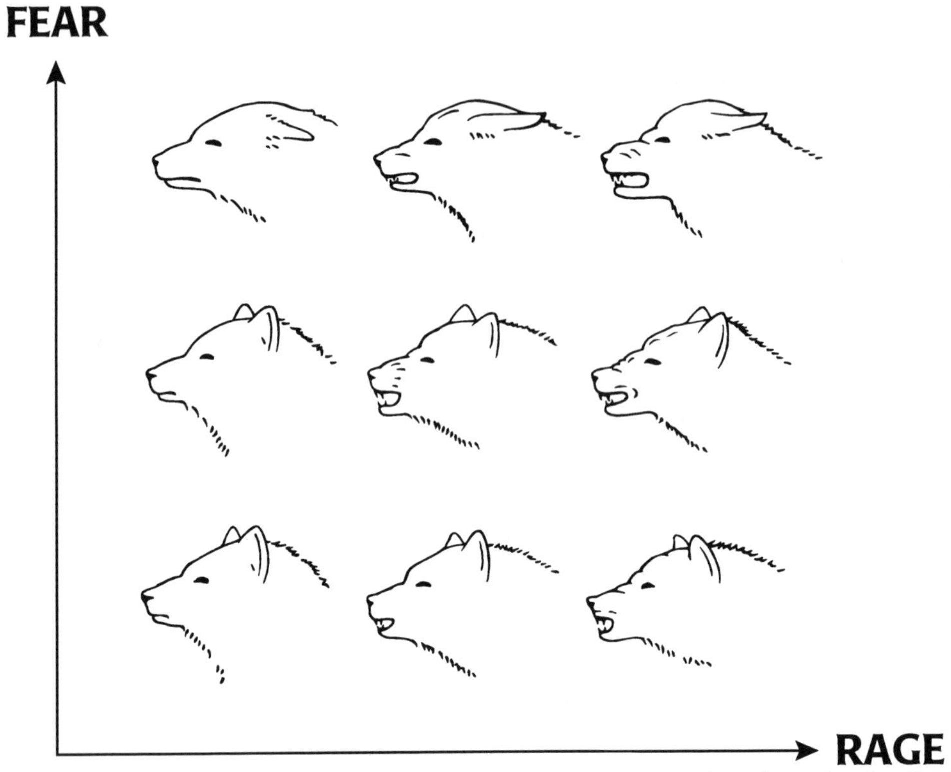

Source: Data from "Catastrophe Theory" by E.C. Zeeman. Copyright © 1976 by Scientific American, Inc. All rights reserved.

1. **a.** Describe what is represented in this graph.
   b. Why is this a "good" example?

Many people do not trust the information provided in statistical charts. Sometimes there is a poor sample, or the data is presented improperly. Some graphs have "mistakes" or "misrepresent" data. When using graphs, it is important to think about how they are made, and to make sure the graphs do not give wrong impressions.

1. a. The pictures are the data points in the coordinate system:

In the lower left corner, the animal is neither angry nor afraid.

In the upper left corner, the animal is not angry but is very afraid.

In the upper right corner, the animal has a high level of anger and fear.

In the lower right corner, the animal is very angry but is not afraid.

b. Answers will vary. One possible answer is that the graph conveys a lot of information, yet it is easy to read.

Overview Students explore a graph that uses pictures instead of numbers and discuss what determines a good data representation.

About the Mathematics This graph does not contain any quantitative information. There is no scaling on the axes. Only ordering is possible: the more to the right, the angrier the dog. You cannot say how much angrier.

Planning You may have students work on problem **1** in small groups.

Did You Know? E. C. Zeeman, author of "Catastrophe Theory" (*Scientific American,* April 1967), noted that the aggression of dogs can be described by a mathematical model that uses the principles of catastrophe theory. The mathematical model is based on the assumption that fear and rage are the controlling factors of aggressive behavior. These factors can be measured by changes in the facial expressions of dogs. Levels of rage are determined by the size of the opening of the dog's mouth. Levels of fear are determined by the degree to which the dog's ears flatten toward the back of the head. The mathematical model can be used to predict a dog's behavior.

The theory states that the neutral state of emotion is found at the origin. If a stimulus increases the dog's rage without introducing fear, aggressive behavior results. At high levels of rage, or if the dog is frightened and a stimulus increases the dog's rage, the model predicts that the dog may suddenly attack. If the dog progressively becomes more fearful while at a high level of rage, the dog will eventually stop the attack and retreat.

If a dog's emotional state is neutral, and a stimulus increases rage and fear simultaneously, the dog will attack or retreat depending on its state of mind before the stimulus was introduced.

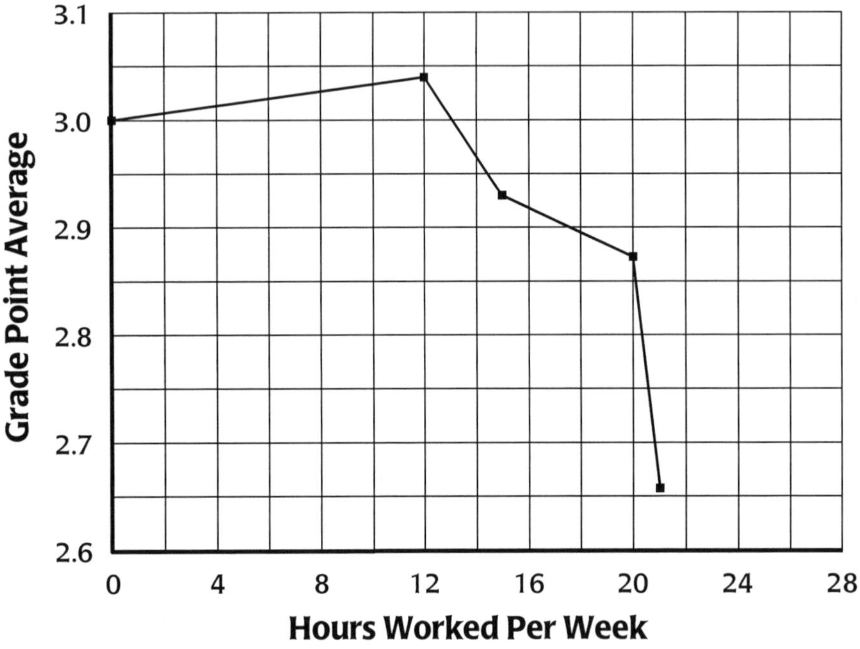

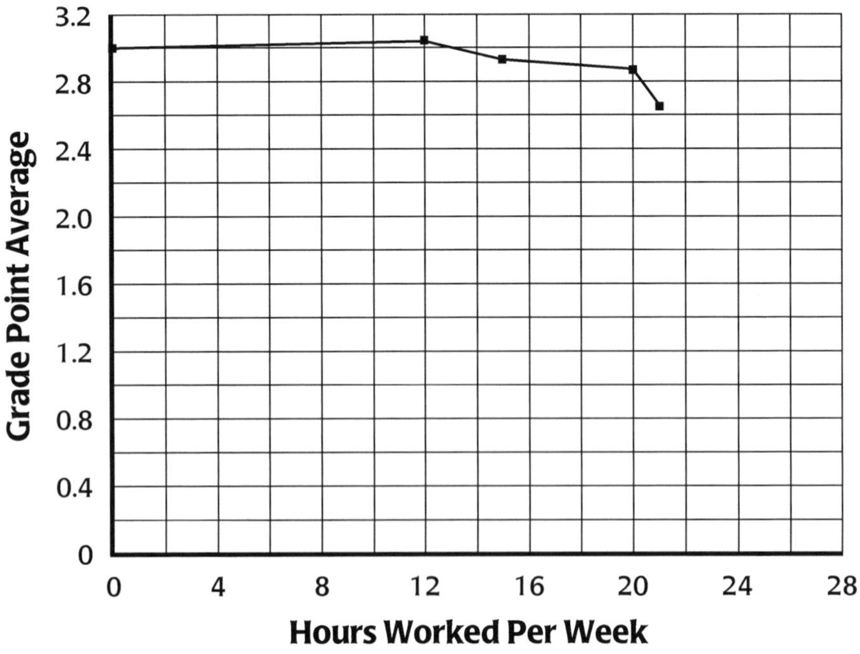

The graphs on this page give two different impressions of the relationship between the number of hours per week that a student works at a job and his or her grade point average.

2. **a.** Why do the graphs provide different impressions?

b. Which graph do you think a high school principal would use when talking to parents about their child's decreasing grade point average? What argument do you think the principal would make?

2. a. Answers will vary. Sample response:

In the first graph, the scale on the vertical axis does not start at zero and is marked in increments of 0.05 for grade point average. The effect of not starting at the origin is to make the drop in grade point average seem very large for students who work more than 12 hours a week.

In the second graph, the scale on the vertical axis is marked in increments of 0.2 for grade point average. And because the scale on the vertical axis begins at zero, the drop in grade point average does not seem as large as in the first graph.

b. Answers will vary. Sample responses:

A high school principal might favor the first graph because it seems to provide an argument against high school students' working long hours on a job.

The first graph is easier to read and is more convincing to support an argument for the reason that working many hours a week is not good for a student.

For someone who believes that working many hours a week does not really affect the grade point average, the second graph is better.

Overview Students investigate the effect of a change in scale and a missing origin in a graph.

About the Mathematics The effect described on this page is known as the *lost origin.* By having the vertical axis start at a value other than 0, the changes are magnified. Usually a break in the axis is used to indicate a missing origin. In some cases, graphs that start at the origin leave much unused space, as in the second graph on Student Book page 22.

Example of a graph with a break in the axis:

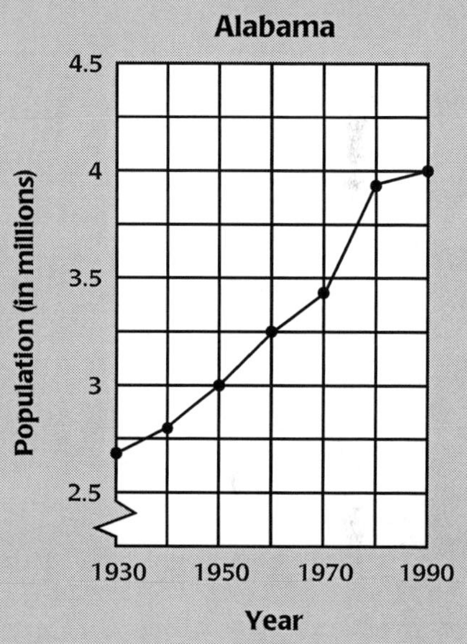

Planning You may want students to work on problem **2** in small groups.

Bringing Math Home You might ask students to begin collecting graphs that distort and misrepresent data. At the end of the section, students may present their collection of graphs to their parents, along with a brief paragraph that explains how each graph misrepresents the data.

Below is a chart of the world population for specific years.

Year	World Population (in billions)
1630	0.5
1820	1.0
1890	1.5
1930	2.0
1950	2.5
1960	3.0
1968	3.5
1975	4.0
1981	4.5
1987	5.0

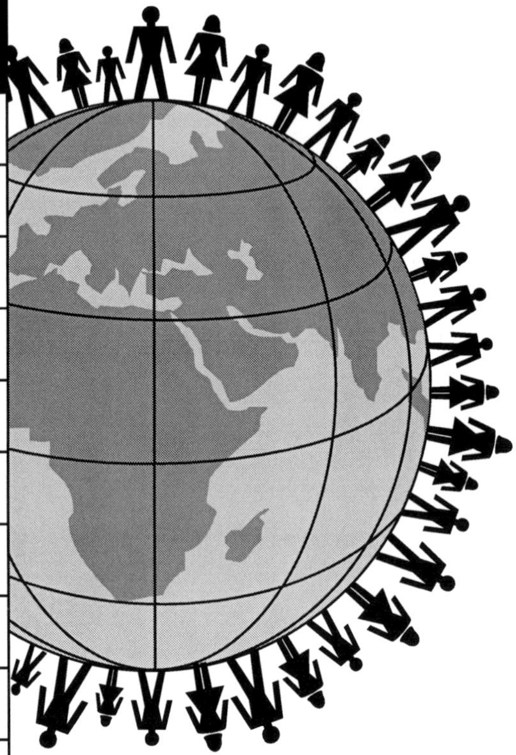

Source: Data from the *World Almanac and Book of Facts 1994* (New Jersey: Funk and Wagnalls Corporation, 1993).

3. Draw a graph to represent the data from the table.

4. a. Based on these data, what do you expect the world population to be in the year 2000?

b. Is it possible to predict the world population in the year 2100? Why or why not?

Use **Student Activity Sheet 5,** which shows examples of graphs made by two students, to answer problem **5.**

5. a. Do these graphs accurately represent the data? Explain.

b. Compare these graphs to the graph you drew in problem **3.**

3. Student's graphs will vary. The following is an example of a line graph.

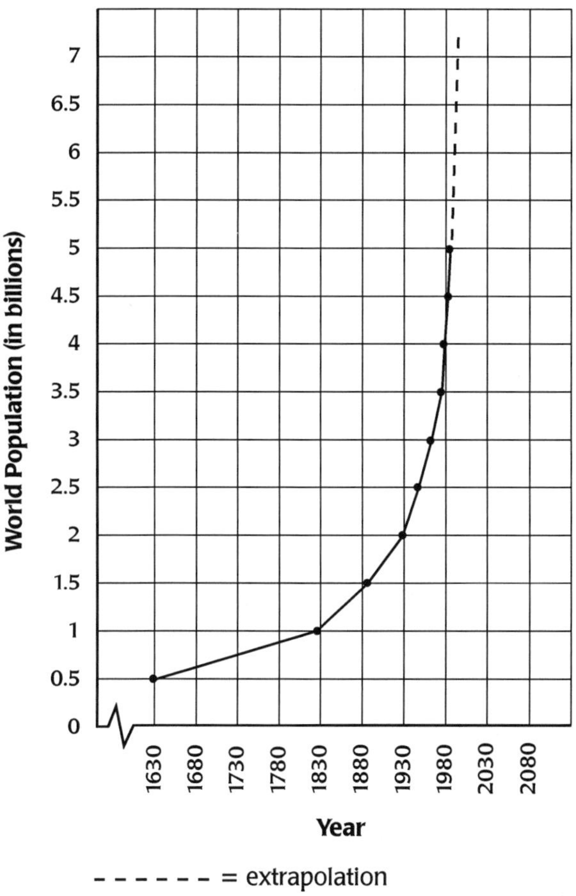

Note: The horizontal axis should not start at 0, and students should have a vertical axis from 0 to about 7 billion.

4. **a.** Answers will vary. A good estimate is about 6 billion people.

   b. Answers will vary. Sample response:

   You can predict the population, but it will be an estimate based on conditions that exist today.

5. **a.** Answers will vary. Sample response:

   No. In both graphs, the intervals on the horizontal axes are not equally spaced. The horizontal axes are scaled incorrectly. Both graphs give the impression that the world population grows in a linear way.

   b. Answers will vary, depending upon students' graphs from problem **3.**

Materials Student Activity Sheet 5 (one per student); transparency of Student Activity Sheet 5, optional (one per class); overhead projector, optional (one per class); graph paper (one sheet per student)

Overview Students make and analyze graphs that represent the world population growth.

About the Mathematics The *lost origin* was the problem on the previous page. Here the scaling of the axis is addressed. In general, the axes should be scaled linearly, meaning that the distances between tic marks are the same.

In most tables (and graphs) that represent growth over time, the units of time are at equal intervals.

Planning Students may work on problems **3–5** in small groups. After students complete problem **5,** you may want to have a class discussion about the lost origin and the scaling of axes. You may find it helpful to make a transparency of Student Activity Sheet 5 to use during class discussion.

Comments about the Problems

3. To make a histogram, students should draw the scale for the horizontal axis correctly, leaving blank spaces for years for which the population is not given in the table. However, a line graph that is correctly scaled on the horizontal axis seems more appropriate to use in this case.

5. **Informal Assessment** This problem assesses students' ability to analyze representations of data and to determine whether representations of data are appropriate.

The graph below was taken from an article in the November 16, 1992 issue of *Newsweek.*

6. a. What impression does the graph give you about the relationship between the number of hours students worked at a part-time job and their grade point averages?

b. How do the scales of the graph affect your impression?

c. Comment on the statement written in the graph.

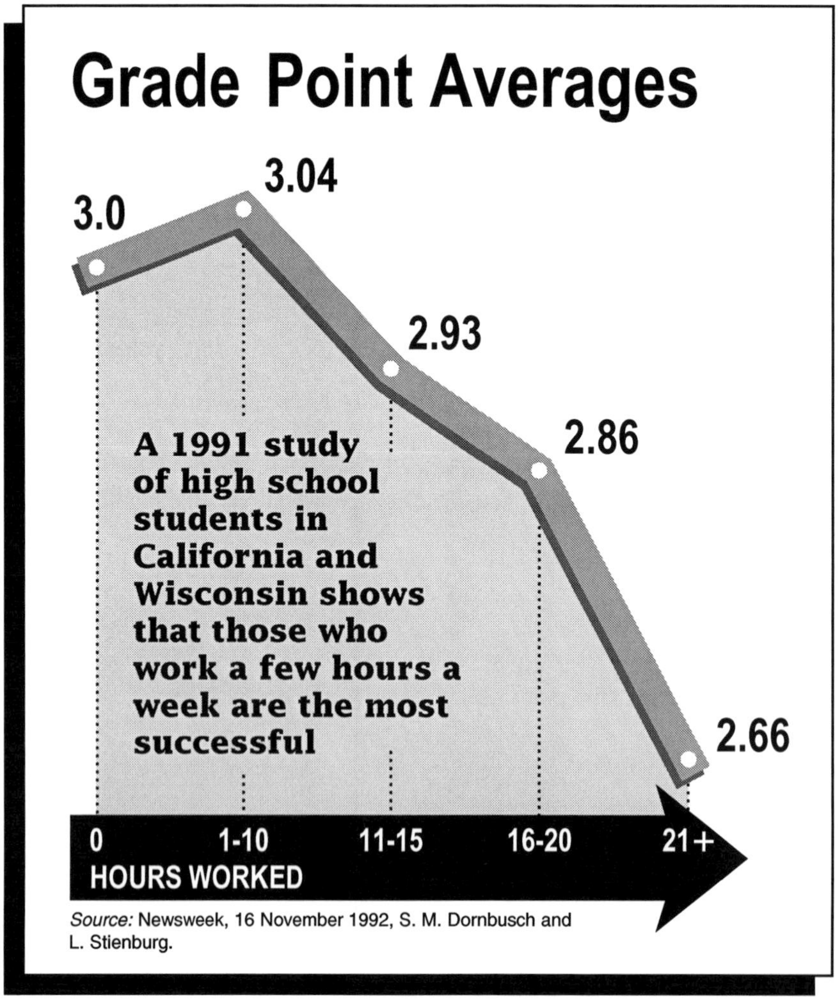

Source: Newsweek, 16 November 1992, S. M. Dornbusch and L. Stienburg.

Solutions and Samples
of student work

6. **a.** The graph gives the impression that the number of hours that students work outside of school has a negative impact on grade point averages.

   b. Answers will vary. Sample response:

   Because the vertical (*y*-axis) does not start at zero, the graph appears to show that grade point averages suffer greatly as students work more hours. The horizontal axis is not scaled in equal time intervals.

   c. Answers will vary. Sample student questions and comments:

   What is meant by "a few hours a week"?

   Are these students who worked 10 hours or those who worked 1 hour? They are in the same group.

   Were these students involved in extracurricular activities also?

   What type of job was considered? Paid jobs? Jobs at home?

   The question of bias is also an issue, since the data were collected from only two states.

Hints and Comments

Overview Students investigate misleading data representation in a graph.

About the Mathematics This graph illustrates a lost origin and the effects of unclear scaling of both axes. Further, there is no information provided about the sample size (that is, the number of students) and the way the sample was taken.

Planning Students may use this graph to demonstrate their improved critical attitude toward data representation. You may want to discuss problem **6** as a class or assign it as homework.

Comments about the Problems

6. **Homework** This problem may be assigned as homework. You may want to ask students how the data were collected. This problem is an opportunity to review sampling, which was the focus of Section B.

Extension You may want to draw the graph again using a consistent scale on the horizontal axis. *Note:* This graph was used as the basis for the graphs on Student Book page 22, although the scaling is slightly different in each of those.

Railway Accidents in the United States

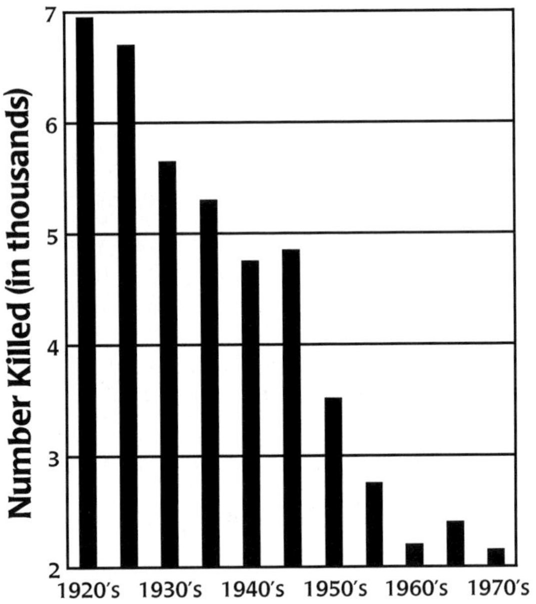

Source: Sunburst Communications

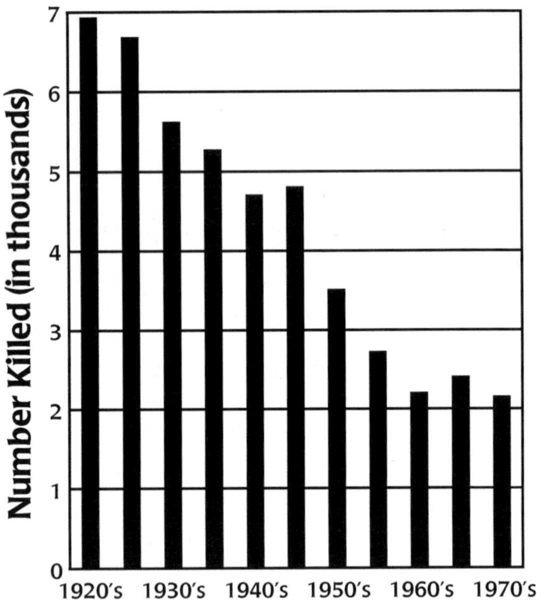

Although the histograms above represent the same data, they give different impressions.

7. a. Which histogram suggests that traveling by train has become very safe?

b. How was this impression achieved?

c. What group of people might use each graph? Why?

d. Do you think the data for these graphs were found by sampling? Give reasons to support your answer.

Solutions and Samples
of student work

7. **a.** The graph on the left suggests that traveling by train has become very safe.

   b. The bars for 1950–1970 are much smaller than the others. The impression is that the number of people killed during these years was minimal. This impression is caused by the fact that the vertical scale does not start at zero.

   c. Answers will vary. Sample responses:

   The president of a railroad might use the graph on the left to indicate that traveling by train has become very safe.

   Individuals promoting train travel might use the graph on the left to give people the impression that the safety of train travel has improved significantly.

   A public action committee that is concerned about the safety of train travel might use the graph on the right, since improved safety may be needed.

   d. Answers will vary. The data for these graphs were likely based on the whole population. Although the numbers have probably been rounded, it is presumable that all deaths from railway accidents in the United States have been recorded.

Hints and Comments

Overview Students investigate two graphs that represent the same data but give different impressions.

Planning Problem **7** can be used as informal assessment.

Comments about the Problems

7. **Informal Assessment** This problem assesses students' ability to describe the relationship between two variables and to analyze representations of data. It also assesses their ability to recognize possible bias in sample surveys and to determine whether representations of data (numerical and visual) are appropriate.

Commission Payments to Travel Agents
(in millions of dollars)

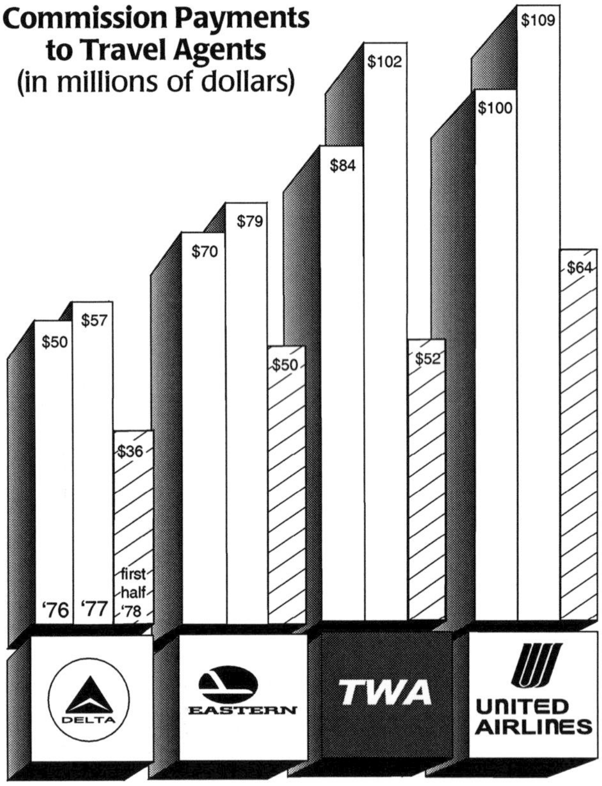

Source: Copyright © 1978 by The New York Times Co. Reprinted by Permission.

8. **a.** Do the commission payments for travel agents appear to be accurate for the year 1978? Why or why not?

   b. Sano drew a graph doubling the amounts shown for 1978. Do you think this is an accurate way to represent the commission for travel agents?

   c. What group of people might use this graph? Why?

Annual Cost per Person for Care of the Mentally Ill in Pittsburgh City Homes and Pennsylvania State Hospitals

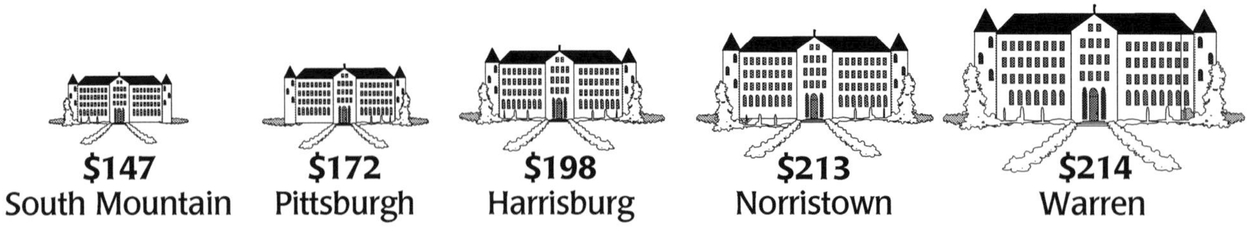

9. **a.** What message does the above graph tell you?

   b. Do you think the graph expresses the message in a fair way? Why or why not?

8. a. Answers will vary. Sample response:

No. The bars for 1978 are misleading. The small text on the bar for Delta 1978 indicates that the amount is for the first half of 1978. The bars for 1976 and 1977 represent commissions for the whole year. Also, if you look at the $50 bar for Eastern in 1978 and the $50 bar for Delta in 1976, you can see that they are different sizes.

b. Answers will vary. Sample response:

Doubling implies that the commissions earned by travel agents would be the same during the second half of the year as in the first half. This does not take into account Thanksgiving and the winter holiday season, the busiest travel times of the year.

c. Answers will vary. Sample response:

Travel agents may use the graph to raise complaints about the amount of commissions they received. The travel agents may request a higher percent rate for their commissions.

9. a. Answers will vary. Sample response:

The graph gives the message that the cost for care in Warren is much higher than in any other city home or state hospital.

b. The graph does not present the information in a fair way.

Explanations will vary. Sample explanation:

No, the graph does not express the message in a fair way. The difference in costs for the two institutions on the right is only one dollar, yet the sizes of the buildings are very different.

Overview Students investigate two graphs that either do not present all the information or create a misleading picture.

About the Mathematics The first graph illustrates how missing or incomplete data can be the reason for a wrong impression. Further, the heights of the bars do not correctly represent the payments. In the second graph, the absolute increase (in dollars) is not directly represented by the increase in area of the picture.

Planning Problems **8** and **9** are optional. If time is a concern, these problems may be omitted or assigned as homework.

Comments about the Problems

8. Not only do the bars for 1978 represent just the first half of the year, but also the scaling of the graph is not consistent. For instance, compare the 1976 bar for Delta to the 1978 bar for Eastern. Both bars represent $50 but are not the same height. Also, the difference between $84 and $102 is drawn only slightly larger than the difference between $100 and $109.

9. Have students compare the increases in dollars with the increases of the pictures. Ask students, *What do you think is more persuasive, the increases in the dollar figures or the increases in the sizes of the buildings?*

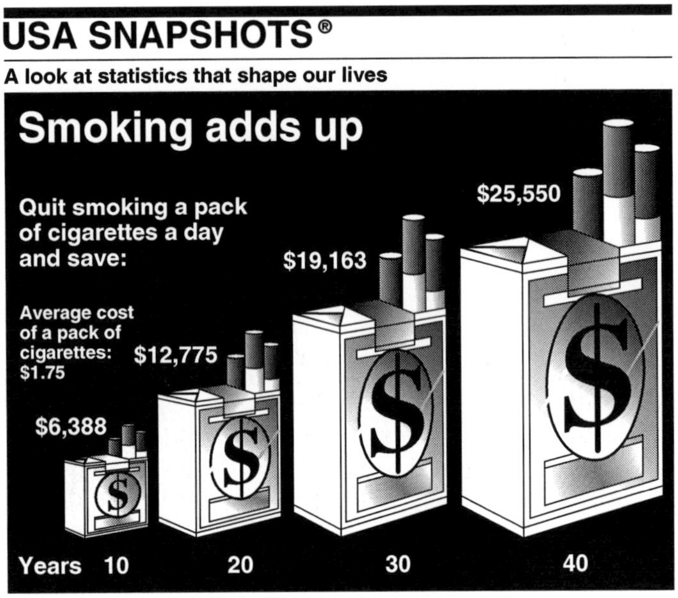

Source: Copyright 1993, USA TODAY. Reprinted with permission.

Look carefully at the graph on the left.

10. a. How was $6,388 computed?

b. What data are used to make this graph?

c. The last package has $25,550 written above it. The amount $25,550 is 4 × $6,388, rounded to the nearest ten. Explain how this fits the data.

d. Is the last package four times the size of the first? Carefully explain your answer.

11. The graph above gives misleading visual impressions since the sizes of the pictures don't match the numbers. Draw a new graph that accurately portrays the data.

Activity

12. Design and carry out a statistical survey. Pay attention to the following aspects:

- What are you going to investigate?
- How do you write a good questionnaire?
- What is a good sample for your survey?
- How will you graphically represent your data?
- What conclusions can you make from your data?
- Do any new questions arise from your data?
- What further investigations might be necessary?
- Is there a possibility of bias in your survey? If so, can you eliminate it?

Solutions and Samples
of student work

10. a. The figure of \$6,388 was calculated by taking the average cost of a pack of cigarettes (\$1.75) and multiplying that cost by 365 days to find the annual cost (\$1.75 × 365 = \$638.75). To find the cost after 10 years, you multiply the annual cost by 10 (\$638.75 × 10 = \$6,387.50).

b. The data used are the average cost of a pack of cigarettes, the number of days in a year, and the number of years.

c. Explanations may vary. Sample response:

Forty is four times ten, the interval of time represented by the first pack of cigarettes. The amount \$25,550 is approximately four times \$6,388. The numbers are accurate. But they can be questioned, since over 40 years the cost of a pack of cigarettes would not remain the same.

d. Answers will vary. Sample student responses:

Yes, the last pack is four times as high as the first.

No, the volume of the last pack is more than four times the volume of the first pack.

11. Graphs will vary. One method would be to draw stacks of blocks. One block represents \$6,388 (the savings for 10 years), a stack of two blocks represents \$12,775 (the savings for 20 years), and so on.

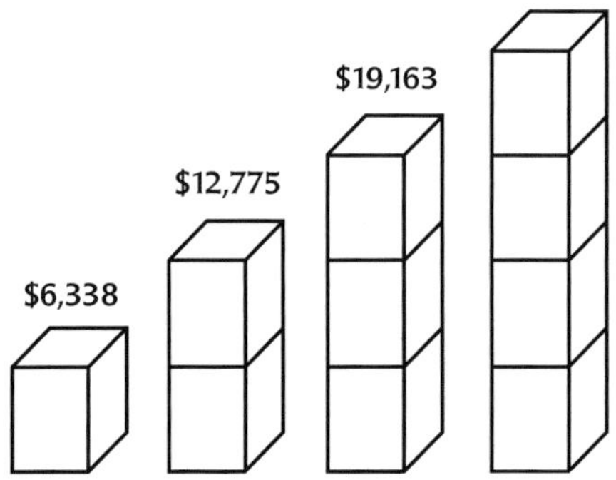

12. Surveys will vary. Students should present their data graphically and write about the conclusions they have drawn.

Hints and Comments

Materials graph paper (one sheet per student)

Overview Students explore the effect of volume in the representation of data.

Planning Have students work on problems **10** and **11** individually. You may want students to work on the activity (problem **12**) in small groups. You might give students one week to complete their surveys. Have students hand in a report and/or present their results to the class. This activity can be used as an informal assessment.

Comments about the Problems

10. d. The impression you get from the graph is that the last pack is much bigger than the first one. Since the volume of the pack gives this visual impression, the graph is misleading.

11. Compare students' drawings with those of the original graph. Discuss the differences between the representations of the data.

12. Informal Assessment This activity assesses students' ability to design, conduct, and analyze ways of gathering data; to use random samples in gathering data; and to recognize possible bias in sample surveys.

You might discuss the process of planning the survey with the class. As they design their surveys, students need to think ahead to the analysis of the data and to how the data will be used to answer questions. They may consider questions such as, What should be measured? How should it be measured? How many individuals should be included in the survey? How should these individuals be selected? It is also important to discuss your expectations for students' work (what is to be included, how should it be presented) and when the survey results are due. Students may also include a self-critique of the study, noting any possible bias.

Writing Opportunity You may want students to write a report that summarizes their experiences in designing and completing their surveys.

Summary

In order to make accurate conclusions from data, the data must be reliable and presented appropriately.

There are many different ways to present data:

- pictorial graphs
- line graphs
- bar graphs
- histograms

Data may be misrepresented if one or more of the following occurs:

- the graph's axes are scaled improperly
- origins on the graph are excluded
- three-dimensional pictures are used inappropriately
- numbers that should not be compared are compared
- pictures that do not fit the numbers are used.

Summary Questions

13. Why is it important for a graph to be an accurate representation of data?

14. Cut out a graph from a newspaper or magazine. Make sure to include the caption or article that accompanies the graph. Attach the graph with the article or caption to a sheet of paper. Then write a paragraph that explains how the graph more clearly presents the information in the caption or article. Do you think the graph is fair or unfair?

13. Answers will vary. Sample response:

When a graph is an accurate representation of data, relationships are revealed and proper conclusions can be drawn.

14. Answers will vary. Sample student response:

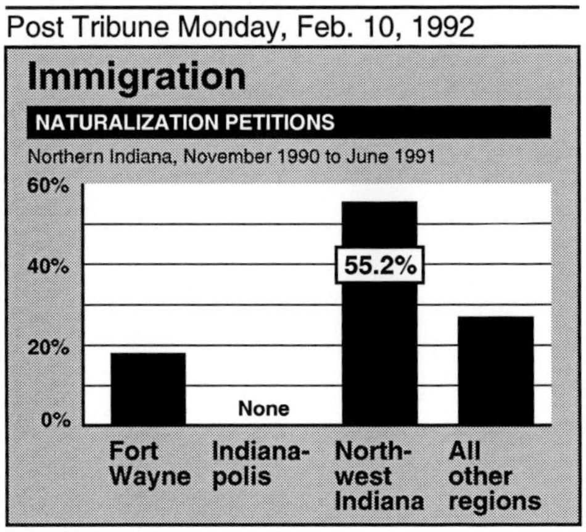

This graph helps the reader to picture what the article talks about. The graph divides northern Indiana into regions and gives the percent of immigrants in each area from November 1990 to June 1991. The reader can relate this information to his or her own community. The graph seems to represent data in a fair way.

Hints and Comments

Materials graphs from newspapers or magazines (one per student); blank paper (one sheet per student); glue or tape (one per student)

Overview Students read the Summary, which reviews the main topics of this section. They then consider possible reasons for graphical misrepresentation and comment on a graph they cut out from a newspaper or magazine.

About the Mathematics The most common types of misrepresentations in graphs are listed in the Summary.

Planning Students may work on problems **13** and **14** individually. It is useful to have some examples of misleading graphs available for problem **14** in case students weren't able to find any. These problems may be used as informal assessment. After students complete Section C, you may assign appropriate activities from the Try This! section, located on pages 58–62 of the *Insights into Data* Student Book, for homework.

Comments about the Problems

13–14. Informal Assessment These problems assess students' ability to analyze representations of data; to determine whether representations of data are appropriate; and to become aware of the questions that should be asked when analyzing data sets and representations of data.

14. Informal Assessment This problem also assesses students' ability to represent data graphically; to design, conduct, and analyze ways of gathering data; and to draw conclusions based on given data sets and representations of data.

Extension You might ask students to compile some of the misleading graphs they collected during this section and write explanations describing how each one misrepresents the data.

Work Students Do

This section emphasizes the use of graphs and numbers to display the results of the bean sprouts experiment that was conducted in Section A. Students analyze their data from the experiment. They select appropriate plots to summarize the data and use histograms and box plots to compare information.

Goals

Students will:

- represent data graphically;
- describe data numerically;
- design, conduct, and analyze ways of gathering data: surveys, simulations, experiments;
- analyze representations of data;
- draw conclusions based on given data sets and representations of data;
- recognize possible bias in sample surveys; *
- determine whether representations of data (numerical and visual) are appropriate;
- become aware of the questions that should be asked when analyzing data sets and representations of data.

** This goal is assessed in other sections of the unit.*

Pacing

- approximately three 45-minute class sessions

Vocabulary

- box plot
- histogram
- plot over time

About the Mathematics

In this section, several plots and graphs are used to represent data about the growth rate of bean sprouts in various solutions. Box plots and histograms were introduced in the unit *Dealing with Data* and used in Section B. Most of the concepts from statistics (measures and graphs) that appear in this section were introduced in the unit *Dealing with Data.*

Students make histograms and box plots that show the lengths of sprouts grown in various solutions. They discuss using a plot over time (usually a line graph) to show the data points as they change from day to day. There are many ways to make a plot over time: plot the growth of a representative member of the population; plot the range in growth per day; plot the mean or median length per day; or plot the growth of each individual in the sample.

A plot over time allows you to look for trends in the data. A histogram best presents clusters and gaps in the data. A box plot provides a summary of the five major data points: the minimum, the first quarter, the median, the third quarter, and the maximum.

Materials

- transparency of Student Book page 29, pages 69 and 71 of the Teacher Guide, optional (one per class)
- overhead projector, pages 69 and 71 of the Teacher Guide, optional (one per class)
- photocopies of Student Book page 29, page 71 of the Teacher Guide, optional (one per student)
- scissors, page 71 of the Teacher Guide, optional (one pair per student)
- graph paper, pages 73 and 79 of the Teacher Guide (several sheets per student)
- centimeter rulers, pages 73 and 81 of the Teacher Guide (one per student)
- completed Student Activity Sheet 1, pages 79 and 81 of the Teacher Guide (one per group of students)
- colored self-stick notes, page 81 of the Teacher Guide (four notes per group of students)

Planning Instruction

You may want to introduce this section with a brief review of statistics and summary measures. You might ask questions such as, *What is a statistic?* [A statistic is a number that summarizes a set of data. It may be a count, a representative value in a set of data, or an arithmetic combination of data points.] *Why do we need summary measures?* [If a data set is very large, summary measures, such as the mean, the median, and the mode, are used to organize and make sense of the data.] You might want to also discuss common statistics in students' lives, such as magazine polls and sports box scores.

Students may work on problems 1–3, 8, and 9 individually. Problems 4–6 can be worked on in small groups, followed by a class discussion. Problem 7 may be done as a whole class.

Problem 6 is optional. If time is a concern, you may omit this problem or assign it as homework.

Homework

Problem 5 (page 78 of the Teacher Guide) may be assigned as homework. After students complete Section D, you may assign appropriate activities from the Try This! section, located on pages 58–62 of the *Insights into Data* Student Book. The Try This! activities reinforce the key mathematical concepts introduced in this section.

Planning Assessment

- Problem 3 can be used to informally assess students' ability to represent data graphically and to analyze representations of data.
- Problem 5 can be used to informally assess students' ability to represent data graphically; describe data numerically; and to design, conduct, and analyze ways of gathering data: surveys, simulations, experiments. It also assesses their ability to analyze representations of data; draw conclusions based on given data sets and representations of data; and become aware of the questions that should be asked when analyzing data sets and representations of data.
- Problem 8 can be used to informally assess students' ability to represent data graphically and to determine whether representations of data (numerical and visual) are appropriate.
- Problem 9 can be used to informally assess students' ability to describe data numerically; determine whether representations of data (numerical and visual) are appropriate; and become aware of the questions that should be asked when analyzing data sets and representations of data.

D. USING PLANT GROWTH DATA

Exploring Growth

Graphs of data will help you get a better picture of the distribution of the data. Graphs can help you interpret the data and make statements about the experiment.

The graph below represents the growth of plants after a seven-day period.

Plant Height at 7 Days

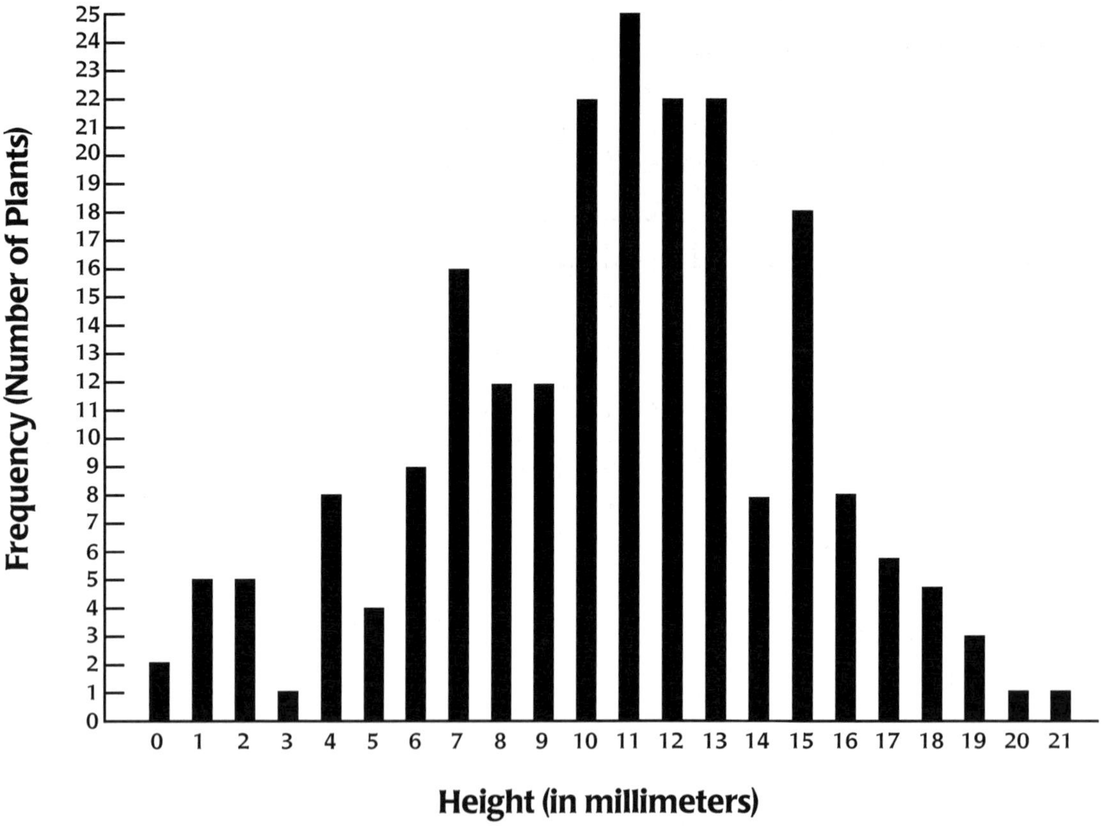

1. a. Study the graph. What does the bar at 7 millimeters represent?

b. Write three statements about the growth of the plants in the study.

Solutions and Samples
of student work

1. a. The bar at 7 millimeters represents the 16 plants that grew to a height of 7 millimeters by the end of seven days.

   b. Answers will vary. Sample student responses:

   The shortest height of any plant that grew was 1 millimeter.

   The tallest height was 21 millimeters.

   Most of the heights were between 4 and 15 millimeters.

   Twenty-five plants grew to a height of 11 millimeters.

   More plants grew to 11 millimeters than to any other height.

   Two hundred and fifteen plants are being studied.

   The mean plant height is about 10.6 millimeters.

Hints and Comments

Materials transparency of Student Book page 29, optional (one per class); overhead projector, optional (one per class)

Overview Students study and interpret a bar graph representing data on plant growth.

About the Mathematics In the bar graph, data on three variables are represented: the heights, the frequency, and the growth period of the plants. For different growth periods, you would need different graphs. Another way to represent growth would be to use a line graph over time. Since so many plants are involved, however, a line graph would be very difficult to read.

Planning Students may work on problem **1** individually. A transparency of the graph may be helpful during a class discussion.

Comments about the Problems

1. Additional questions can be used to determine whether or not students interpreted the graph properly. For instance:
   - What is the shortest plant height?
   - How many plants grew to that height?
   - What is the tallest plant height?
   - What is the range of the most common plant heights?

   b. Students may use the mean and the mode in their statements; these concepts have been taught in earlier units. The mode is the most frequent data point. In this case, it is the height that has the tallest bar (11 millimeters).

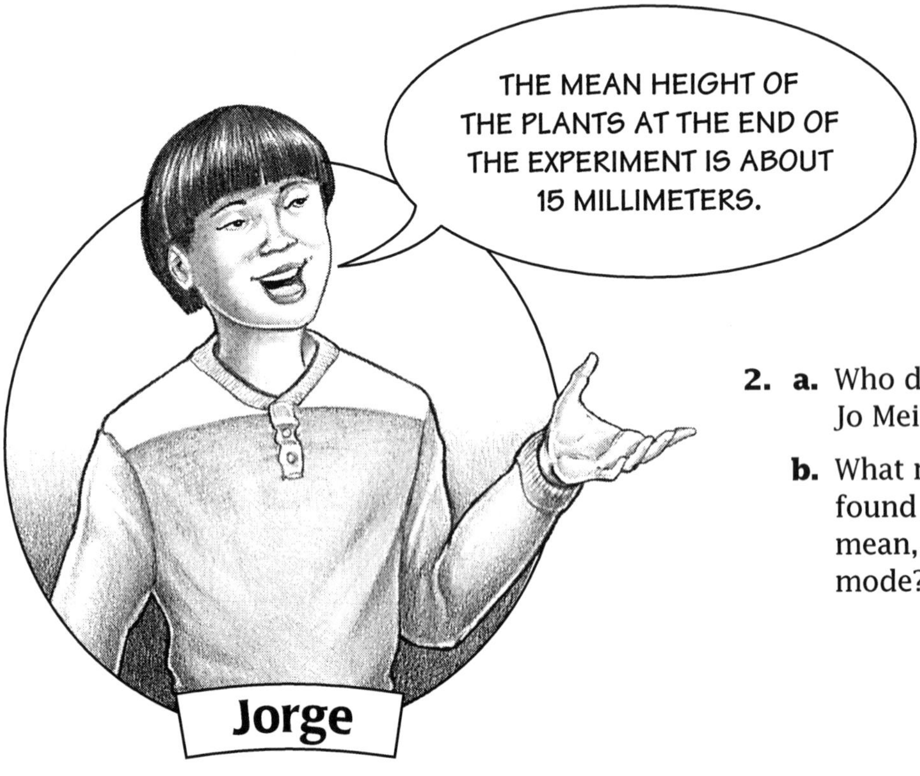

2. **a.** Who do you think is right, Jo Mei or Jorge?

b. What number is most easily found in the graph: the mean, the median, or the mode?

Solutions and Samples
of student work

2. a. Jo Mei's estimate seems to be closer to the mean plant height, which turns out to be about 10.6 millimeters. Jorge's estimate of 15 millimeters is too high for the mean.

   b. The mode is most easily found in the graph. The most frequent plant height is 11 millimeters.

Hints and Comments

Materials transparency of Student Book page 29, optional (one per class); overhead projector, optional (one per class); photocopies of Student Book page 29, optional (one per student); scissors, optional (one pair per student)

Overview Students connect statements about the plant data to the data on the graph. The one-number summaries of the mean, the mode, and the median are revisited.

About the Mathematics In *Dealing with Data,* students learned that the mean is a "balance point" of the distribution for the histogram. If a histogram is cut out of paper and a fulcrum (such as the length of a pencil) is placed at that point, the distribution will balance.

Planning Students may work individually on problem **2.**

Comments about the Problems

2. a. One way to find the mean height is by reasoning from data on the graph: The total height to the right of 15 millimeters is much less than the total height to the left of 15 millimeters. This can be estimated by multiplying *height* × *frequency* for all heights. It is possible to calculate the exact mean, but estimations are sufficient at this point.

   b. The median is the data point in the middle of the distribution. The following method can be used to find the median: First count to find the total number of plants in the study (215). Then count from left or right till you find plant number 108. This plant has the median length. It is in the bar with plants of 11 millimeters, so 11 millimeters is the median height.

Akir, Kari, Viviana, and Marja are studying the growth of plants. After seven days, they each graphed their data. Then they wrote statements about the growth of their plants. Unfortunately, some of the work was misplaced, and the rest was mixed together. Below is what is left of their work.

3. For each student, match the statement with a graph. If a student's graph is missing, make an appropriate graph.

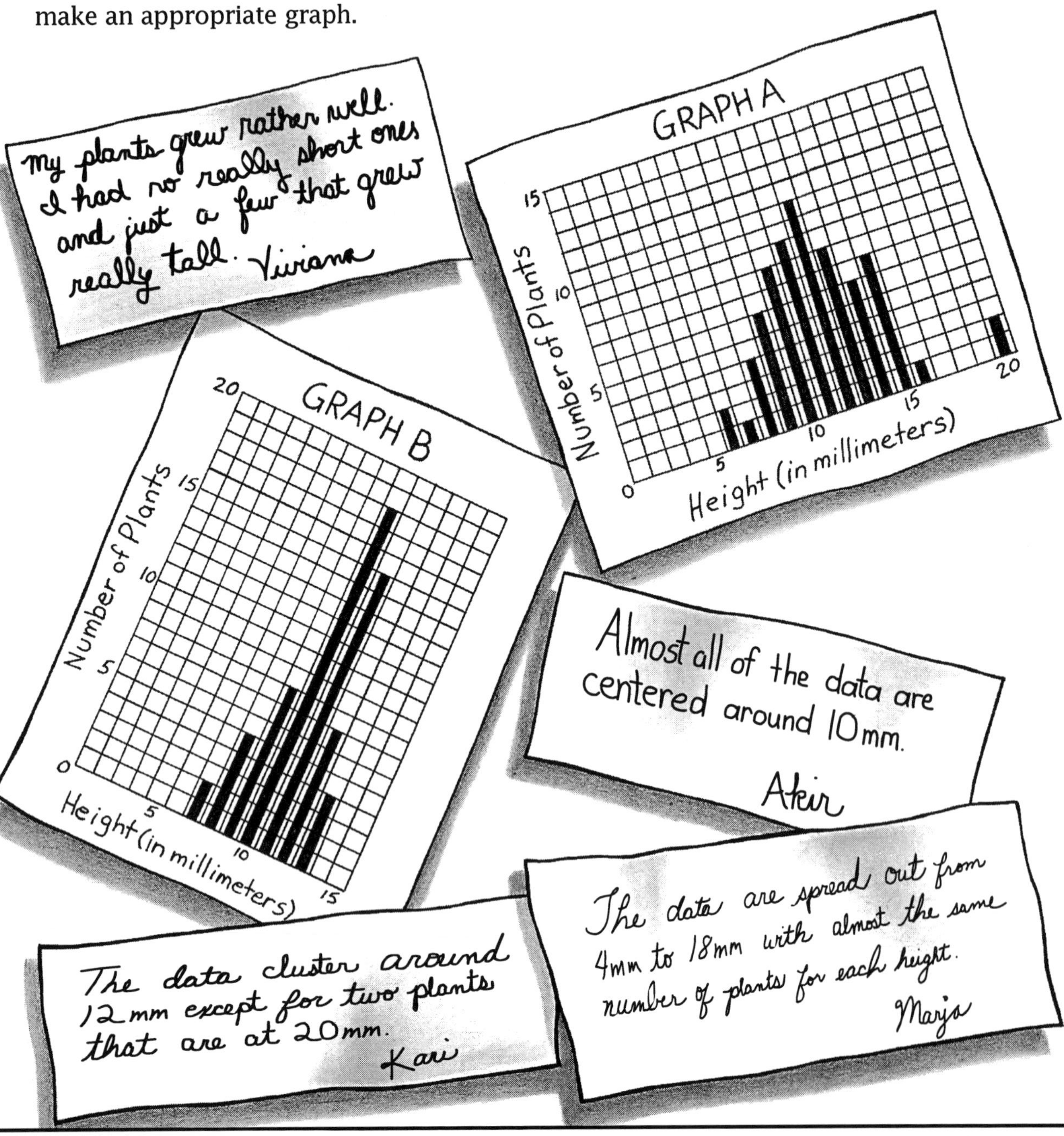

3. Students must make appropriate graphs to represent the data for Viviana and Marja. Possible graphs are shown below.

Kari—"The data cluster around 12 mm, except for two plants that are at 20 mm."

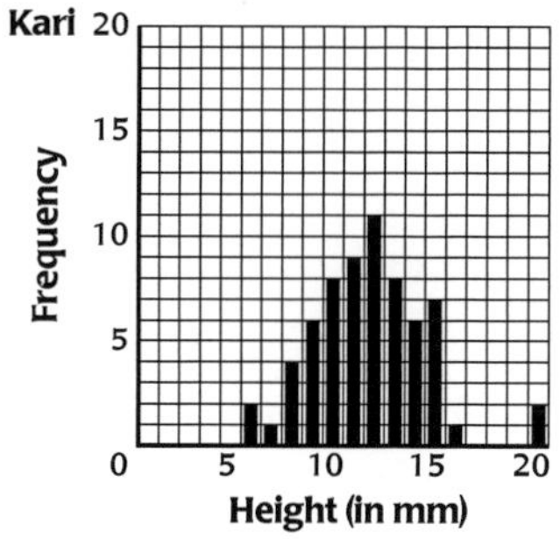

Akir —"Almost all of the data are centered around 10 mm."

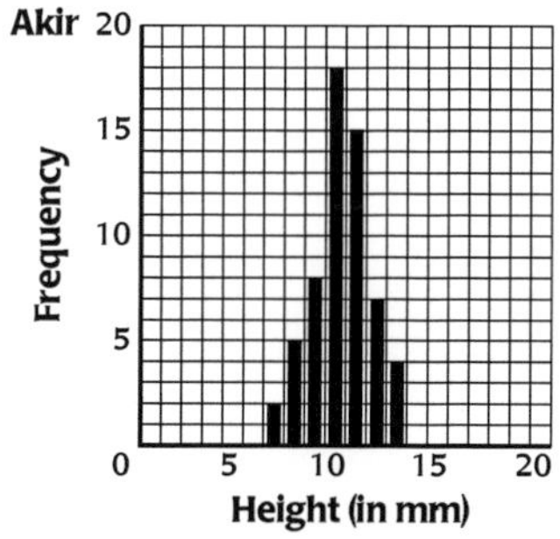

Viviana—"My plants grew rather well. I had no really short ones and just a few that grew really tall."

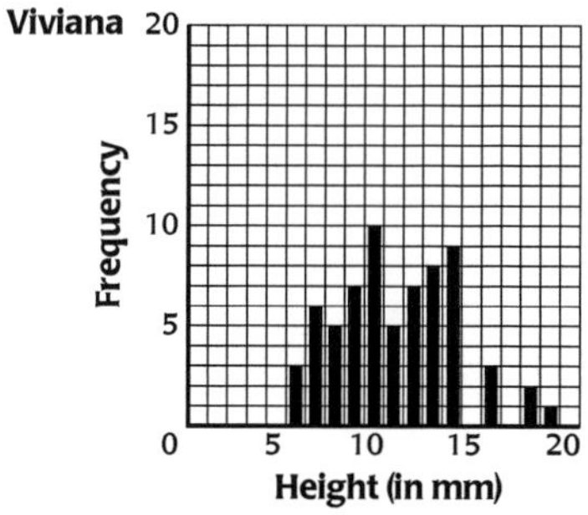

Marja—"The data are spread out from 4 mm to 18 mm with almost the same number of plants for each height."

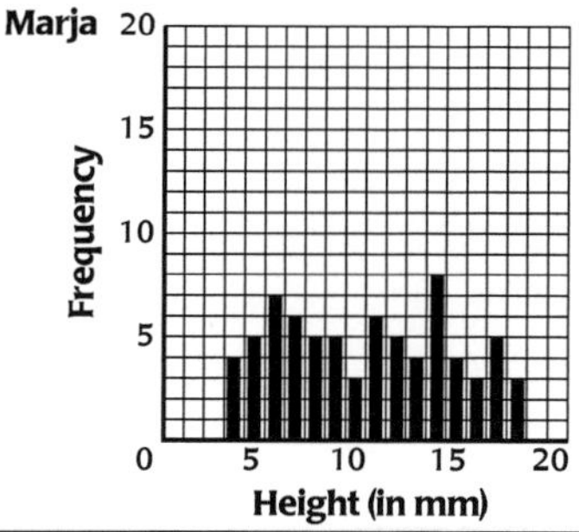

Hints and Comments

Materials graph paper (several sheets per student); centimeter rulers (one per student)

Overview Students connect statements to graphs and vice versa.

About the Mathematics Understanding the relationship between statements and graphs is an important ability. Problems about connecting graphs and statements can also be found in the unit *Ups and Downs.*

Planning Students may work on problem **3** individually. This problem may be used as informal assessment.

Comments about the Problems

3. **Informal Assessment** This problem assesses students' ability to represent data graphically and to analyze representations of data. Other solutions may be possible. It can be argued that graph A also fits the statements of Viviana and Akir. If students choose one of these options, make sure they correctly graph the remaining statements.

The histograms shown below and on the next page represent the frequency of heights of a group of plants at 10 days, 12 days, and 14 days.

4. a. Explain what each histogram tells you about the plant heights.

b. Describe the growth pattern of these plants.

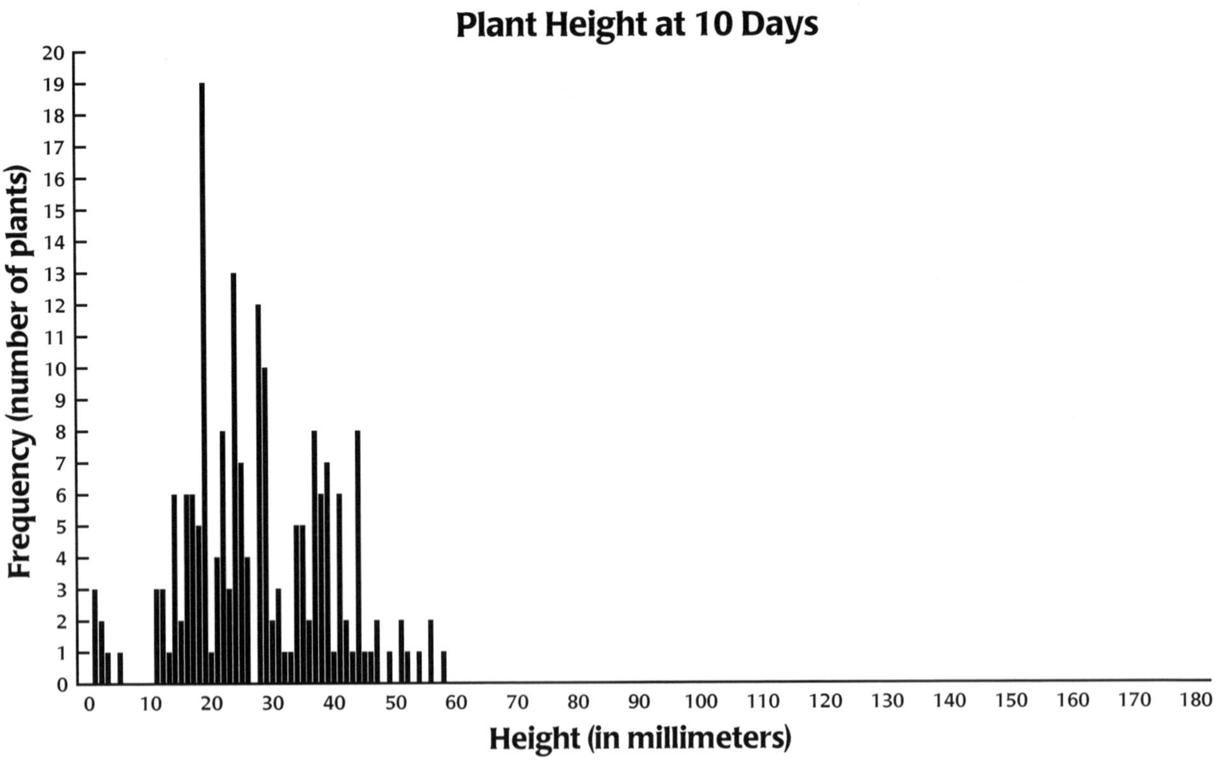

Solutions and Samples
of student work

4. a. Answers will vary. Sample response:

Plant height at 10 days: The heights of the plants varied. Some of the plants did not grow much at all, while others grew to a height of 58 millimeters. Most of the plants grew to between 20 and 30 millimeters.

Plant height at 12 days: The heights of the plants were more spread out than at 10 days and ranged from 0 millimeter to 107 millimeters. Most of the plants grew to heights of between 30 and 70 millimeters.

Plant height at 14 days: The heights ranged from nearly 0 millimeter to 159 millimeters, with one outlier that grew to a height of 180 millimeters. With a few exceptions, namely, at 25 millimeters, 85 millimeters, 110 millimeters, 125 millimeters, and 150 millimeters, there were one or two plants at many heights.

b. Answers will vary. Sample response:

With time, the heights of the plants become more varied. After 10 days, the plants ranged in height from about 10 to 50 millimeters, with most plants at around 30 millimeters. After 12 days, most of the plants were between 30 and 70 millimeters tall, and after 14 days most of the plants were between 50 and 140 millimeters tall.

Hints and Comments

Overview Students interpret and compare three histograms showing the heights of plants after different growth periods.

About the Mathematics The growth of a large group of plants can be presented in a series of histograms as shown on these pages. If the scaling of the axis is the same for all histograms, statements can be made and conclusions can be drawn about the growth process of the whole group. The growth of individual plants cannot be tracked.

Planning Students may work in small groups on problem **4.** Be sure to discuss their answers before proceeding to the next problem.

Comments about the Problems

4. b. If students are having difficulty, find out whether they are correctly interpreting the graphs. Students may interpret the heights of the bars as the heights of the plants.

Plant Height at 12 Days

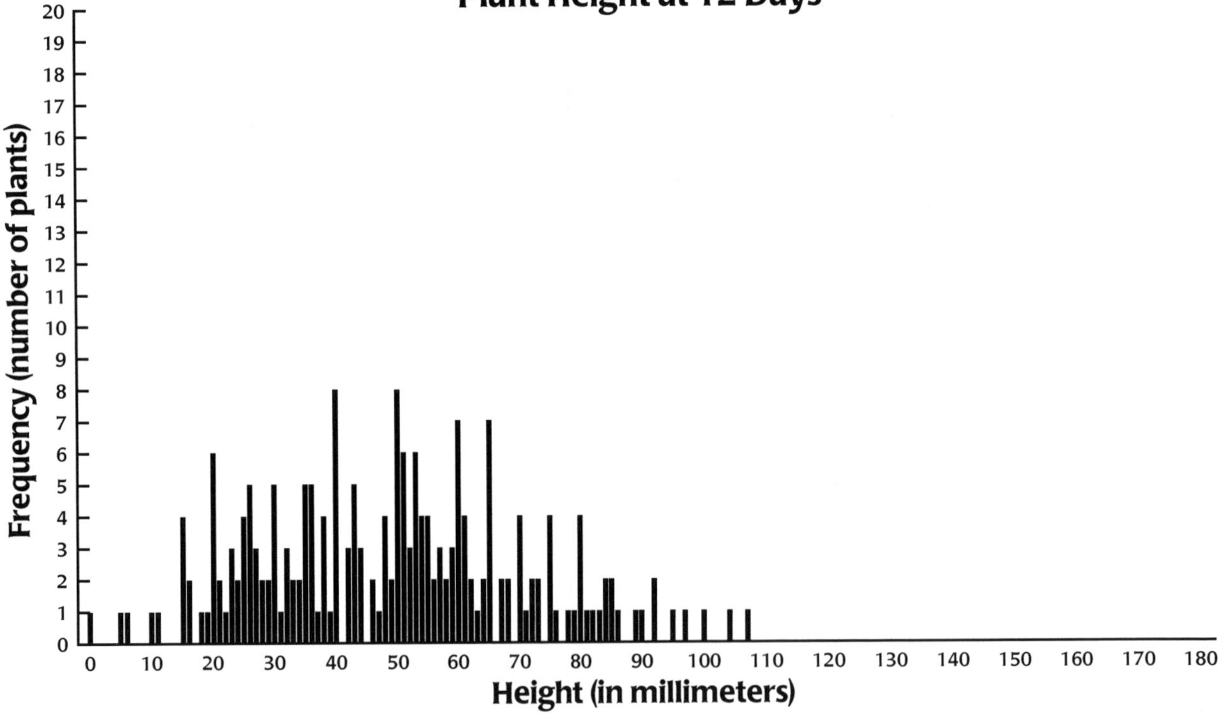

Plant Height at 14 Days

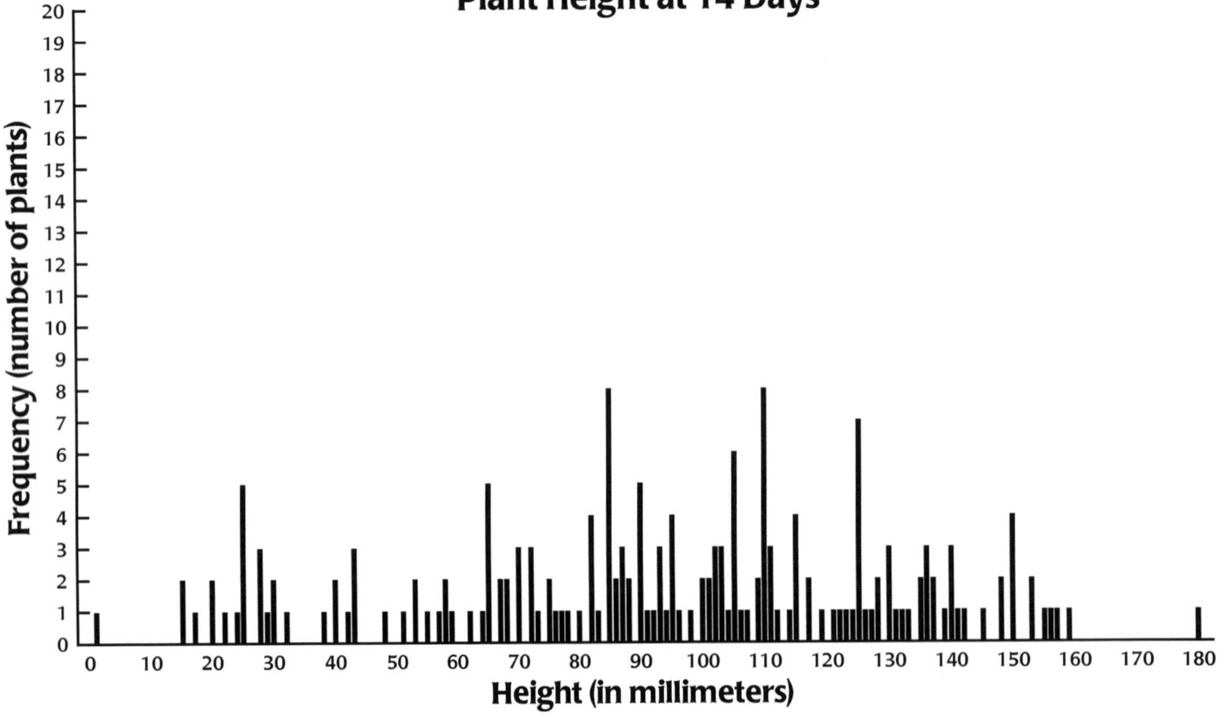

Overview Histograms for the problem on the previous page are shown here. There are no additional problems on this page for students to solve.

Presenting the Bean Sprout Data

In the bean sprout growth experiment you conducted in Section A, you collected data and kept a record of the lengths of the bean sprouts and your observations.

5. Prepare a report for the class on the results of your experiment. Your report should include the following:

- the data you collected,
- a graph of the final length of each sprout,
- a written description of what the graph shows about the final sprout lengths,
- a graph of the growth of the sprouts over time,
- a written description of what the graph shows about the growth of the sprouts over time,
- some statements about the data in which you use the mean, median, or mode (select the one you feel is best for your data),
- a statement about how the growth of the beans varied,
- your conclusions about how the bean sprouts grew in the solution you used,
- a list of the things that affected the results of your experiment,
- a list of the things you would change if you were to repeat the experiment.

Solutions and Samples
of student work

5. Reports will vary.

The final lengths will vary with the solution. For example, in one experiment, the final lengths of bean sprouts grown in tap water were 44, 57, 95, 95, 85, 70, 45, 75, 40, and 90 millimeters. These data illustrate that, in some instances, data are distributed in such a way that there is no apparent center of the data. This fact is obscured by the box plot, but is revealed by the stem-and-leaf plot, where there is a cluster of lengths at the low end and a cluster at the high end, with few lengths in the middle. Students may select various ways to plot the final lengths: a number-line plot, a box plot, and a stem-and-leaf plot are some examples. Students also may choose a variety of ways to plot the lengths of the sprouts over time. For example, they may use several graphs or use different colors.

The mode does not give significant information about the growth of the beans over time. In one given data set, there may be no mode, one mode, or more than one mode. The mean and median give some indication of the center of the data. The mode doesn't have to be at the center.

Hints and Comments

Materials graph paper (several sheets per student); completed Student Activity Sheet 1 (one per group of students)

Overview Students write a report summarizing the experiment they conducted in Section A. They use statistical tools, such as graphs and numerical measures, to describe their results.

About the Mathematics Statistical tools are often used to describe the results of an experiment. It is important to be able to use these tools correctly.

Planning Students may work in small groups on problem **5.** Students may write their reports during class or you might assign them as homework.

Comments about the Problems

5. Informal Assessment This problem assesses students' ability to represent data graphically; to describe data numerically; and to design, conduct, and analyze ways of gathering data. It also assesses their ability to analyze representations of data; to draw conclusions based on given data sets and representations of data; and to become aware of the questions that should be asked when analyzing data.

The design of an experiment is an important statistical process. In this experiment, the beans were the subjects. Water was the treatment for the control group. The other treatments were the different solutions.

Students might discuss whether the experiment contained bias and whether they could design the experiment in another way to reduce the bias that may have resulted.

Writing Opportunity/Interdisciplinary Connection The bean-growth experiment reports are an excellent way for students to work on their writing skills. After students have completed their reports, you might want each group of students to present their results to the class as well. Students could apply techniques they have learned from language arts, speech, or English classes and work as a group to organize and present their reports.

6. Your teacher will give you some self-stick notes. Write the final length of each bean sprout on a separate self-stick note. Put your notes on the number line your teacher has drawn on the board to make a histogram. Use the finished histogram on the board to answer the following questions:

   a. What seems to be the average length of the bean sprouts?

   b. How did the bean sprouts in your solution compare to the bean sprouts in other solutions? What conclusions can you draw about the effects of your solution on the way the bean sprouts grew?

The histogram gives a general picture of the growth of the bean sprouts in each solution. Suppose you want to compare the effects of the different solutions more directly. One way to compare solutions is with box plots.

7. **a.** Make a box plot for the final lengths of the bean sprouts in your solution. As a class, decide on a common scale for the box plots. Why is a common scale necessary?

   b. Below your box plot, write the type of solution your group used. Put your box plot on the board together with those from other groups. How do the box plots help you compare the different solutions?

   c. What is the difference between the information you can see on a histogram and the information you can see in a box plot?

6. **a.** Answers will vary. Sample response:

The average length of the bean sprouts can be estimated from the histogram by looking at the pattern of the final sprout lengths and approximating the balance point.

b. Answers will vary. Sample response (considering the results of a sample experiment):

The sprouts grown in tap water grew the most. Sprouts grown in cola and lemon-lime soda grew more than those grown in a salt solution, but less than those grown in tap water. The sprouts grown in a salt solution grew the least.

7. **a–b.** Answers will vary. Sample response:

Using the results from a sample experiment, the box plots for the final lengths of the bean sprouts are as follows.

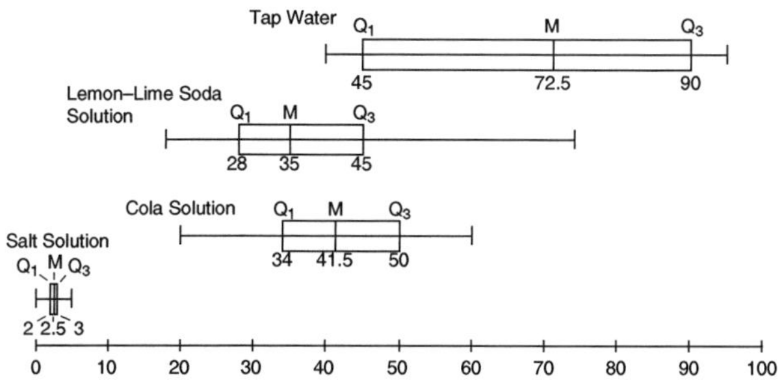

Box Plots for Final Length of Sprouts in Various Solutions

By using a common scale, the information from the box plots can be compared. For example, the box representing the middle 50% of the final sprout lengths in one solution can be compared with the boxes of other plots.

c. Answers will vary. Sample response:

A histogram provides a general picture of the set of data, yet preserves individual data. You can detect gaps or clusters in the data. A box plot provides a summary of the data and can be used to compare sets of data. Box plots also show how consistent the growth was. If the plot is more compact, the growth is more consistent. However, individual values are lost.

Hints and Comments

Materials completed Student Activity Sheet 1 (one per group of students); centimeter rulers (one per student); colored self-stick notes (four notes per group of students)

Overview Groups compare the growth of their beans by means of a histogram and box plots.

About the Mathematics Box plots are very useful when comparing data from different groups, which is the case here. Individual data, however, are lost in a box plot. Since there are only very few data in each group, it is possible to compare the individual data (for instance, in a number-line plot or a stem-and-leaf plot).

Students learned how to draw a box plot in *Dealing with Data.* To draw a box plot, five major data points are needed: the minimum, the first quarter, the median, the third quarter, and the maximum. The box indicates the lengths of the middle 50% of the sprouts. A suitable scale will probably be from 0 to 10 units beyond the length of the longest sprout.

Planning Problem **6** is optional. If time is a concern, this problem may be omitted or assigned as homework. You might want to work on problem **7** as a whole class.

Comments about the Problems

6. A different color of self-stick note should be used for each of the four solutions. When deciding on the length and scale of the number line, take into account the size of the self-stick notes and the range of the data (the shortest and longest lengths recorded).

7. **c.** Box plots offer a quick way to see how growth was affected by each solution. The ends of the whiskers mark the range of growth. If the whisker on one end is long, there was at least one sprout that showed inconsistent growth compared to the others in the set. This is called an outlier.

Summary

A data set can be described numerically by using the mean, median and mode.

A plot over time shows the data points as they change from day to day. The plot allows you to look for trends in the data.

A histogram is a general picture of the data. It allows you to see clusters and gaps in the data.

A box plot gives you a summary of the five major data points: the minimum point, the first quarter, the median, the third quarter, and the maximum point. It shows the big picture of the data but does not allow you to see small details.

Summary Questions

8. When is it helpful to use box plots?

9. Can you estimate the mean, median, or mode from each graph listed below? Explain.

- a histogram
- a box plot

Solutions and Samples
of student work

8. Answers will vary. Students should note that it is helpful to use box plots when comparing groups of data or if you want a summary of the key points in the data.

9. Histogram: The mode is easily seen since it is represented by the tallest bar. The mean can be approximated by looking at the pattern in the shape of the graph. The median can be determined by finding the halfway point. For example, if there are 15 values, the eighth one represents the median.

Box plot: In drawing the box plot, the median is easily recognized as the middle value. There is no way to find the mode or the mean.

Hints and Comments

Overview Students reflect on the statistical tools used in this section.

About the Mathematics Again, the relationship between graphs and numerical measures of data is the issue here. By graphing data or by summarizing the data with a single value, you lose information. It is important to know what information is lost and what information can be retrieved. For example, in a box plot, the median and the range of the data can easily be found, but the mean and the mode are lost.

Planning Students may work individually on problems **8** and **9.** These problems may be used as informal assessments. After students complete Section D, you may assign appropriate activities from the Try This! section, located on pages 58–62 of the *Insights into Data* Student Book, for homework.

Comments about the Problems

8. Informal Assessment This problem assesses students' ability to represent data using box plots and to determine whether representations of data (numerical and visual) are appropriate.

9. Informal Assessment This problem assesses students' ability to describe data numerically; to determine whether representations of data (numerical and visual) are appropriate; and to become aware of the questions that should be asked when analyzing data sets and representations of data.

If the horizontal axis of a histogram is not scaled in single units, you cannot determine the mode by inspecting the histogram. You can determine a category for the mode if one exists.

Work Students Do

In this section, students use scatter plots to investigate the relationship between two variables. They also look for linear patterns in scatter plots. Students learn about correlations, which can be weak, strong, positive, or negative. Students draw conclusions based on scatter plots and correlations and learn that a relationship between two variables is not necessarily a cause-effect relationship.

Goals

Students will:

- describe the relationship between two variables;
- identify the degree of correlation between variables;
- analyze representations of data;
- draw conclusions based on given data sets and representations of data;
- become aware of the questions that should be asked when analyzing data sets and representations of data.*

** This goal is assessed in other sections of the unit.*

Pacing

- approximately one or two 45-minute class sessions

Vocabulary

- cause-effect relationship
- correlation

About the Mathematics

A scatter plot can be used to describe the relationship between two quantitative variables. *Correlation* is a measure of the linear association between the two variables. If the cloud of points in a scatter plot slopes upward and to the right, there is a positive correlation between the variables. If the cloud of points slopes downward and to the right, there is a negative correlation between the variables. The existence of a correlation between two variables, however, does not mean that a cause-effect relationship exists. This section examines the visual perception of linear patterns in scatter plots.

Students are not expected to learn how to compute the strength of the correlation (called the correlation coefficient) in this unit. An intuitive understanding of correlation is sufficient. In Section F, students' intuitive understanding is extended by including straight lines to describe a linear relationship.

Materials

- Student Activity Sheet 2, page 144 of the Teacher Guide, optional (one per student)

Planning Instruction

You may want to introduce this section with a brief discussion about correlation and cause-and-effect relationships. Students should be able to contribute many examples of strong cause-effect relationships. You may want to discuss that reasoning alone is sometimes enough to recognize a relationship, but that at other times, more data is needed.

Students may work on problems 1–5 in small groups. You may want students to work individually on problems 6, 7, 13–15. Problems 8–12 may be completed in small groups or individually.

Problem 13 is optional. If time is a concern, this problem may be omitted or assigned as homework.

Homework

Problems 6 and 7 (page 94 of the Teacher Guide) may be assigned as homework. The Extension (page 97 of the Teacher Guide) may also be assigned as homework. After students complete Section E, you may assign appropriate activities from the Try This! section, located on pages 58–62 of the *Insights into Data* Student Book. The Try This! activities reinforce the key mathematical concepts introduced in this section.

Planning Assessment

- Problems 3 and 5 can be used to informally assess students' ability to analyze representations of data.
- Problem 6 can be used to informally assess students' ability to describe the relationship between two variables and draw conclusions based on given data sets and representations of data.
- Problems 13 and 14 can be used to informally assess students' ability to describe the relationship between two variables and identify the degree of correlation between variables.

E. CORRELATING DATA

Growing Babies

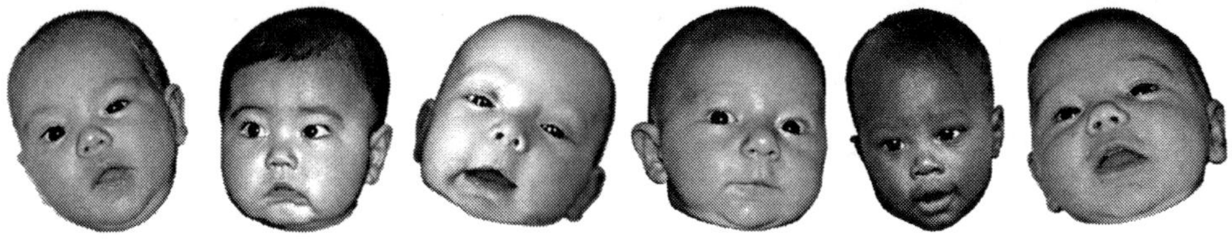

In 1962, researchers studied 100 newborn babies to see whether there was a relationship between the length of the body and the circumference of the head. The table below shows the results of this study.

Circumference of the Head (in centimeters)	Body Length (in centimeters)									
	47	48	49	50	51	52	53	54	55	56
39							|			|
38							|	|		
37					|	|	|	|		
36				|	||||	𝍸 ||	||			
35	|||		𝍸	𝍸 ||||	𝍸 |	|	|			
34	|	𝍸 ||	𝍸 𝍸	𝍸 ||||	|||	|||				
33	|	𝍸 |	𝍸	||||			|			
32		|	|	|						

Source: Sunburst Communications

Most tallies appear along the diagonal that extends from the lower left to the upper right corner of the graph.

1. In relation to the diagonal, where do you find the babies with large heads?

2. As the body lengths increase, what happens to the head sizes of the babies?

Solutions and Samples
of student work

1. Above the diagonal. The data for babies with large heads in relation to their lengths can be found on the upper left side of the table.

2. As the baby's length increases, the head circumference also increases.

Hints and Comments

Overview Students interpret general patterns in a table.

About the Mathematics The illustration shows that the distinction between a graph and a table is not always very clear. The table used here can almost be interpreted as a scatter plot. However, the table depicts *three* variables: body length, head circumference, and the frequency with which certain combinations of length and circumference occur.

Planning Students may work on problems **1** and **2** in small groups.

Comments about the Problems

1. You might ask the following questions to help students who have problems understanding the meaning of the data in the table:

   - *How many babies had body lengths of 52 centimeters?* [12 babies]
   - *How many babies with a head circumference of 34 centimeters had a body length of 50 centimeters?* [nine babies]
   - *What is the most frequent combination of body length and head circumference in the table?* [A combination of a body length of 49 centimeters and a head circumference of 34 centimeters occurs 10 times.]

   Note: Students may notice that the horizontal scale is above the table instead of below, which is more common in graphs.

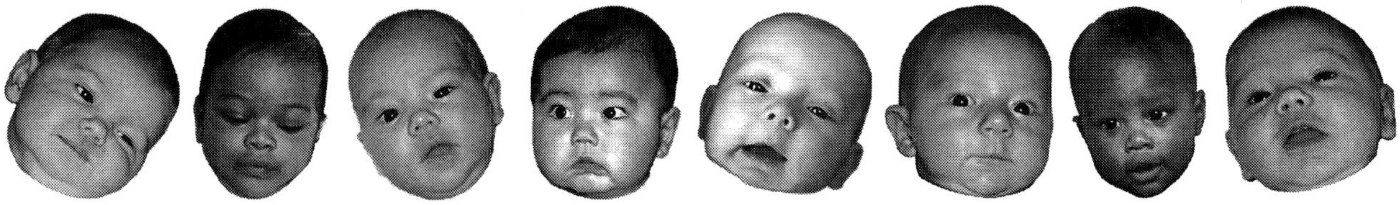

The scatter plot below displays a picture of the information in the table.

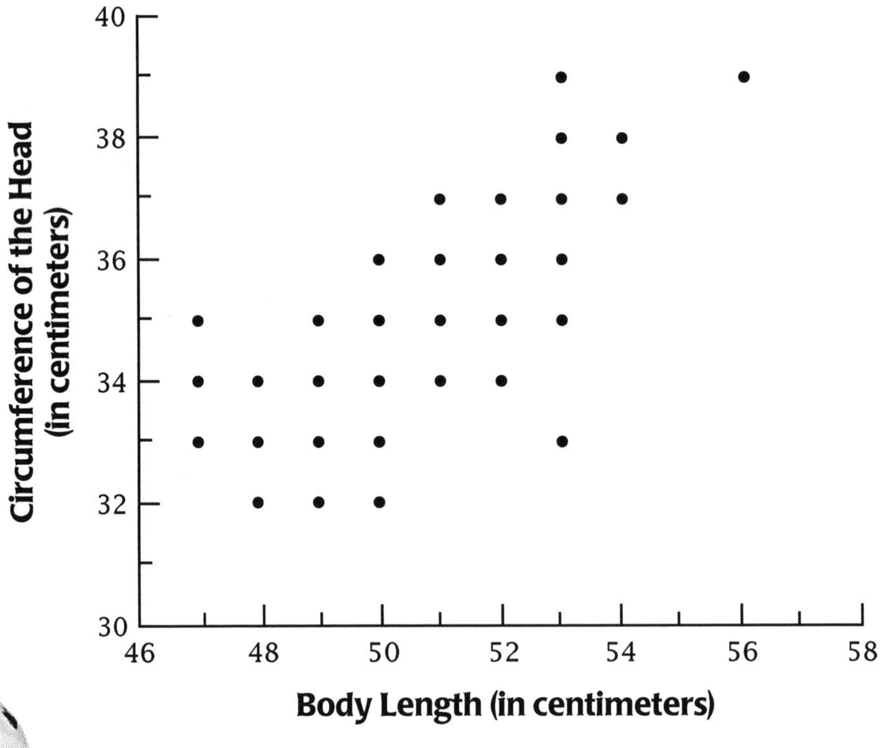

3. What information seems to get lost in the scatter plot when compared with the table?

4. Are there babies for whom body lengths are almost equal to head sizes?

3. In the scatter plot, it is not clear which points represent more than one baby.

4. No. There are no babies who have the same head circumference and body length.

Overview Students compare the table on Student Book page 37 with a scatter plot of the same data.

About the Mathematics In a scatter plot, two variables can be depicted. Sometimes a small number is used beside a point to indicate that this point represents more than one occurrence of a certain combination of variables. But in this example, information about which points represent more than one baby is lost.

Planning Students may work on problems **3** and **4** in small groups. Problem **3** may be used as an informal assessment.

Comments about the Problems

3. **Informal Assessment** This problem assesses students' ability to analyze representations of data.

4. Suggest that students carefully examine the numbers along both axes of the graph. Refer to Section C where the scaling of the axes and the lost origin were discussed. A scatter plot drawn with both axes starting at the same number (for instance, 30) and having the same scale will clearly show that head circumference is significantly smaller than body length.

Extension You might ask students to indicate which babies seem to be different from the others and why. Students might suggest that the baby with a body length of 53 centimeters and a head circumference of 33 centimeters has a small head compared to other babies of that length.

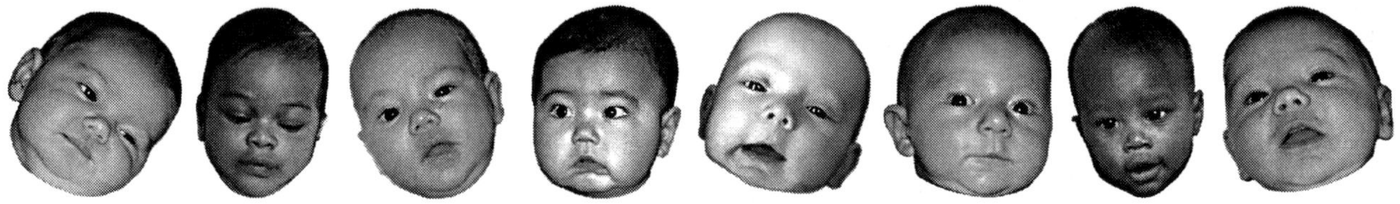

To show body lengths, head circumferences, and the number of babies studied, a three-dimensional graph can be drawn, as shown below.

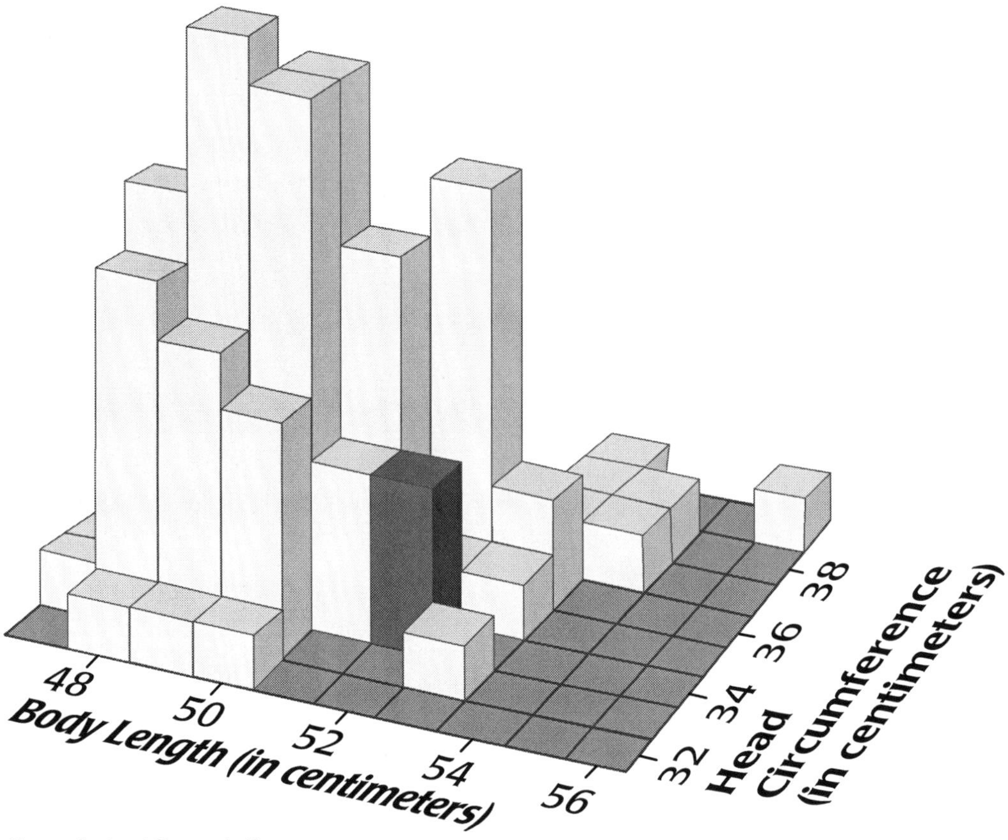

Source: Sunburst Communications

5. a. What information does the shaded bar above represent? Check this information against the data in the table.

b. What are the advantages and disadvantages of a three-dimensional graph?

5. **a.** The shaded bar represents three babies, each having a body length of 52 centimeters and a head circumference of 34 centimeters.

   b. Answers will vary. Sample response:

   The advantages are that the graph preserves most of the information and offers visual impressions.

   The disadvantages are that the graph is difficult to interpret and some bars may be hidden from view by taller bars.

Overview Students read information from a three-dimensional graph.

About the Mathematics Three-dimensional graphs can be used to graph data on three variables simultaneously. This graph is a three-dimensional histogram.

Planning Students may work on problem **5** in small groups. After students have completed problem **5,** you may want to discuss the different representations of data covered thus far.

Comments about the Problems

5. **Informal Assessment** This problem assesses students' ability to analyze representations of data. Students might make connections to top views, side views, hidden views, and so forth, from their work in the geometry strand units. You might ask students to make a top view of the graph using the coding system they may recall from *Side Seeing.*

Juan and Brenda have a discussion about the data.

Hints and Comments

Overview Students read a cartoon in which Juan and Brenda discuss the relationship between a baby's body length and the circumference of its head. There are no problems on this page for students to solve.

6. Do you agree with Brenda or Juan? Explain.

7. a. What do you think the head circumference is for a baby with a body length of 55 centimeters? How did you determine your answer?

b. A one-year-old child has a height of 92 centimeters. What can you tell about the circumference of the child's head?

Solutions and Samples
of student work

6. Answers will vary. Sample student responses:

I can see a pattern, so I agree with Juan. But Brenda makes some very good points to think about.

I think Brenda is right when she says that the relationship will get weaker as the babies grow. As we get taller, our heads don't seem to get that much bigger.

7. a. Estimates will vary. Sample estimate:

The head circumference would be about 38.5 centimeters. One method is to draw a diagonal line as shown on the scatter plot below (known as the line of best fit). This line can be used to estimate the head circumference.

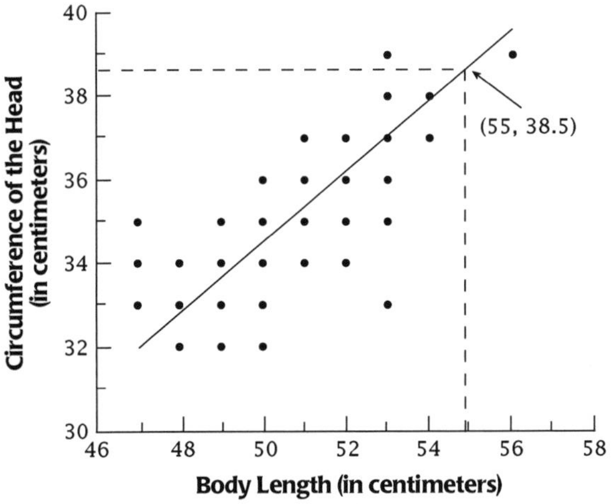

b. Answers will vary. Sample student responses:

The table or graph does not provide this information, so you cannot tell anything about it.

From the graph, I can tell that the baby's head circumference should be greater than 40 centimeters. Since I don't know how the relationship between a baby's length and head circumference continues after a length of 58 centimeters, I cannot approximate the baby's head circumference.

Hints and Comments

Overview Students interpret the data and describe the general pattern of the relationship between a baby's body length and the circumference of its head.

About the Mathematics Looking for a general pattern in the scatter plot informally introduces the concepts of correlation between two variables and the line of best fit. This ties in with the activities connected to the scatter plot in Section A.

Planning Problems **6** and **7** may be assigned as homework. Discuss students' answers briefly in class.

Comments about the Problems

6–7. Homework These problems may be assigned as homework.

6. Informal Assessment This problem assesses students' ability to describe the relationship between two variables and to draw conclusions based on given data sets and representations of data.

To help students understand the scatter plot, suggest that they copy it and write numbers next to the data points, indicating the number of babies represented by each point.

Individual data can detract from the general pattern without actually disturbing it. The relationship is moderately strong and could get weaker. As a person grows, his or her body length changes more rapidly than head circumference.

7. a. Students can imagine or draw a line of best fit and use it to estimate the best-fitting head circumference. The line of best fit is introduced in Section F.

b. Since the data were collected from newborn babies, the relationship between the variables may not be the same for older children.

Scatter plots are often made to show the relationship between two variables. The points on a scatter plot, or scatter diagram, look like a "cloud" of points. When there is a strong relationship between the two variables, the graph resembles a line. If the points appear as a straight line, you can say there is a strong *correlation* between the two variables.

8. Describe the correlation in the following scatter plots. Indicate whether there is no correlation, a weak correlation, or a strong correlation.

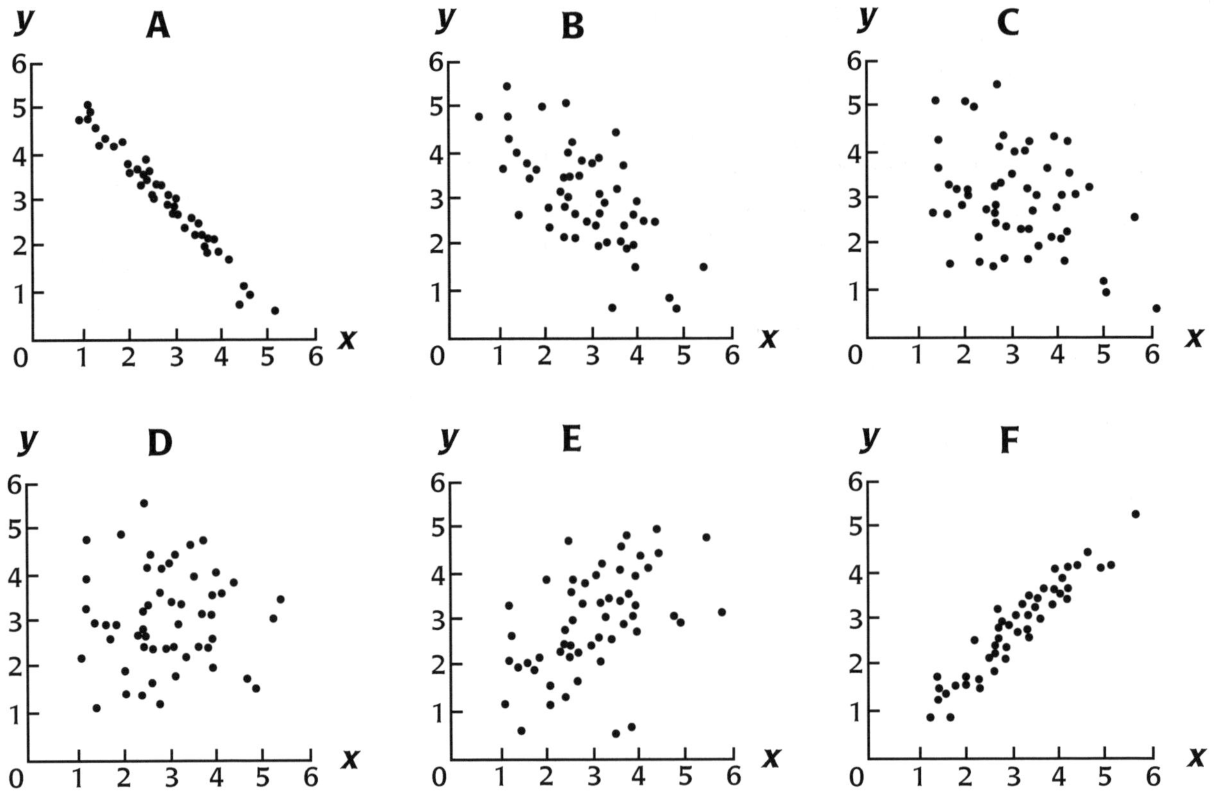

9. For which of the above scatter plots can you best predict the value of *y* when $x = 4$? Explain your reasoning.

In the diagrams above, you can see that a correlation can be weak or strong. Correlations can also be positive or negative.

10. a. What do you think is meant by the phrase "negative correlation"?

b. Which of the above scatter plots show a negative correlation?

Solutions and Samples
of student work

8. Scatter plots C and D show no correlation.

Scatter plots B and E show a weak correlation.

Scatter plots A and F show a strong correlation.

9. You can best predict the value of y when $x = 4$ in scatter plot A because the points form a tight cloud that resembles a straight line. To make a prediction of the value of y when $x = 4$, draw one line vertically from 4 to the cloud of points and another line from the cloud left to the vertical axis, as shown below. A possible value for y is 1.5 or 2.

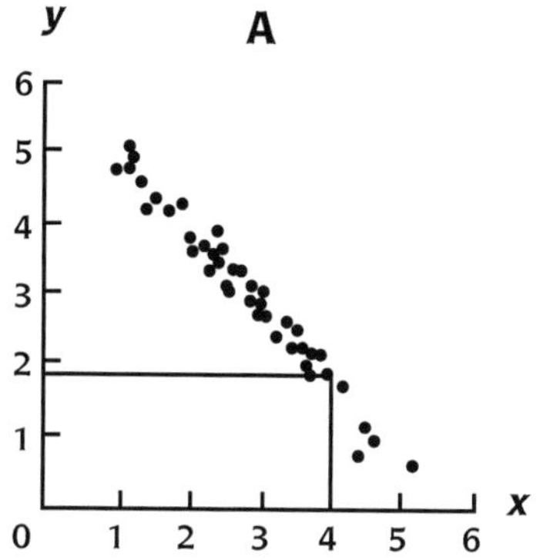

In scatter plot F, the prediction will be less accurate because the points don't form as tight a cloud.

10. a. Answers may vary. A negative correlation exists if the values of one variable increase while the values of the other decrease. In a negative correlation, the cloud of points in a scatter plot will slant from the upper left to the lower right.

Using similar reasoning, a positive correlation exists if the values of one variable increase while the values of the other also increase. In a positive correlation, the cloud of points will slant from the lower left to the upper right.

b. Scatter plots A and B show a negative correlation.

Hints and Comments

Materials Student Activity Sheet 2, optional (one per student)

Overview Students learn the meaning of *correlation* and how the strength of it can be read from a scatter plot.

About the Mathematics This unit introduces a rather informal notion of linear correlation. The correlation coefficient, which is a measure for the strength of the correlation, is not introduced in this unit.

Planning Students may work individually or in small groups on problems **8–10.** After they have completed these problems, see whether they know what is meant by strong, weak, positive, and negative correlations.

Comments about the Problems

8. There are many gradations between a weak correlation and a strong correlation. Students might say that scatter plots B and E show a moderately strong correlation, closer to strong than to weak.

9. In all graphs, you can find points for which x exactly equals 4, and so find the corresponding y-value, but that is not the focus of this problem. Students should use the shape of the cloud, or an imaginary line that summarizes it, to estimate what y-value best fits a data point with an x-value of 4.

Extension You may want students to refer to the scatter plot on Student Activity Sheet 2 from Section A and ask them, *What is the correlation between the percent of urban population by state and per capita income?* [The scatter plot shows a moderate to weak correlation between percent of urban population and per capita income.]

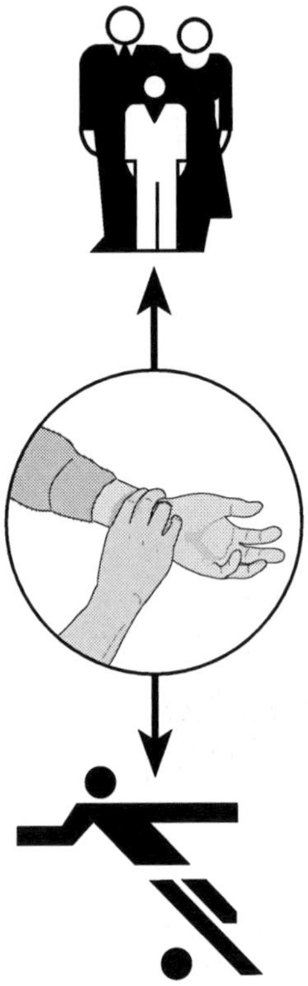

11. For each of the following cases, decide whether there is no correlation, a strong correlation, or a weak correlation between the two variables mentioned. If there is a correlation, is it positive or negative?

- a person's height and pulse rate
- the number of hours of sports training per week and pulse rate
- the height of a dinosaur and the length of its tail
- results on a math test and a science test
- temperature outside in the summer and kilowatt hours of electricity used
- number of children per household and number of television sets per household
- number of hours students study and their grade point averages

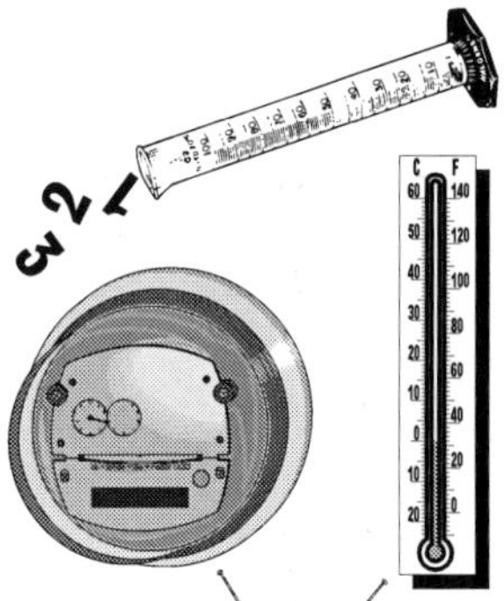

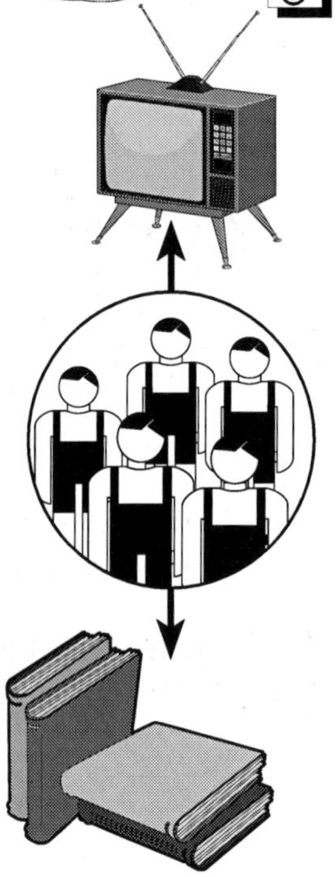

In a *cause-effect relationship* between two variables, a change in one variable directly causes a change in the other.

12. a. If you expect a strong correlation between the two variables, can you be sure that a cause-effect relationship exists? Why or why not?

b. Do you think there is a cause-effect relationship in any of the situations in problem **11?**

11. Answers will vary. Students may offer different perspectives that are valid. For some of the examples, students can carry out an experiment or survey. Sample responses:

- a person's height and pulse rate: no correlation
- the number of hours of sports training per week and pulse rate: negative, weak correlation
- the height of a dinosaur and the length of its tail: positive, strong correlation.
- results on a math test and a science test: positive, weak correlation
- temperature outside and kilowatt hours of electricity used: positive, strong correlation
- the number of children per household and the number of television sets per household: no correlation
- the number of hours students study and their grade point averages: positive, strong correlation

12. a. No. Even though there may be a strong correlation between two variables, you cannot be sure that one variable actually causes the other. It is reasonable to expect that sports training does affect pulse rate. But while grades on math and science tests may be a function of attendance or effort, understanding one subject does not cause one to understand the other.

b. Answers will vary. Sample response:

Yes, the temperature outside in the summer and the kilowatt hours of electricity used have a cause-effect relationship.

Overview Students decide what kind of correlation will exist between variables in different situations.

About the Mathematics Correlation between variables does not mean there is a cause-effect relationship between them. Sometimes a third variable can be found that explains the relationship. A famous example, taken from *A Mathematician Reads the Newspaper* by John A. Paulos (New York: Basic Books, 1995): children with longer arms seem to perform better on reading tasks. Here it is obvious that arm length does not cause better reading. What causes both is the age of the child.

Planning Students may work individually or in small groups on problems **11** and **12.**

Comments about the Problems

11. If students are having difficulty formulating correlations in general terms, encourage them to express the correlations in their own words.

12. It is important that students understand that a strong correlation does not necessarily imply the existence of a cause-effect relationship.

Interdisciplinary Connection Students may ask the physical education teacher or the biology teacher for information about some of the cases described in problem **11,** such as the relationship between training time and pulse rate. However, the explanation of the presence or absence of correlation should remain the primary focus.

Activity

13. Collect some data about one of the cases in problem **11** and use the data to check your answer about the correlation. Be sure to think carefully about how you will select your sample.

Summary

A scatter plot shows information about pairs of data and whether a relationship exists between them. By looking for a trend in the scatter plot, you can see the relationship between the pairs of data.

If the points on a scatter plot are close to forming a straight line, there is a strong correlation. If the points are scattered all over and do not follow a pattern, there is a weak correlation.

A scatter plot with a strong correlation does not imply that there is a cause-effect relationship between the variables. Other tests must be done to determine whether a change in one variable causes a change in the other.

Summary Questions

14. Explain how a scatter plot can help you understand correlation.

15. Find an example of two variables that have a strong correlation but do not have a cause-effect relationship.

13. Answers will vary. Sample response:

The results of investigating the relationship between a person's height and pulse rate are as follows.

Height/Pulse	Height/Pulse	Height/Pulse
5'7"/40	5'6"/45	5'2"/50
5'5"/55	5'4"/58	5'7"/60
5'2"/60	5'1"/60	4'8"/60
5'5"/65	5'4"/66	5'8"/70
5'6"/70	5'0"/70	5'7"/72
5'3"/72	5'2"/75	5'6"/78
5'6"/78	5'3"/92	5'1"/95
5'4"/100	5'3"/103	4'9"/105

There is no correlation.

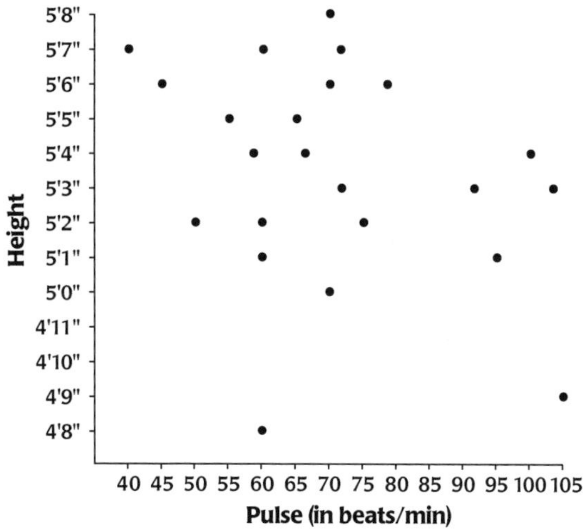

14. Explanations will vary. Sample explanation:

If the points in the cloud form a tight straight line, there is a strong correlation. If the points are scattered widely in the cloud, there is no correlation.

15. Answers will vary. Sample response:

People who spend a lot of money on clothes pay more for a haircut than most people do.

Overview Students collect data for the activity and read the Summary. Then they reflect on the relationship between scatter plots and correlation.

Planning Problems **13–15** may be done individually. These problems may also be used as informal assessment. After students complete Section E, you may assign appropriate activities from the Try This! section, located on pages 58–62 of the *Insights into Data* Student Book, for homework.

Comments about the Problems

13–14. Informal Assessment These problems assess students' ability to describe the relationship between two variables and to identify the degree of correlation between variables.

14. One classic example of a strong correlation between two phenomena without a causal relationship is the number of storks found and the number of babies born in Denmark. Statistical data on a regional level shows a strong correlation between the two, where obviously there cannot be a causal relationship.

Work Students Do

Students study the relationship between the lengths and widths of bird eggs by making scatter plots for various families of birds. They draw lines of best fit on the scatter plots in order to predict widths for given lengths or vice versa. This leads to an exploration of the slope of the line of best fit. The least squares linear regression line is then introduced, and students use a graphing calculator to find the equation for the line. Students also explore the median fit line and compare it with the least squares linear regression line.

Goals

Students will:

- represent data graphically;
- describe data numerically;
- describe the relationship between two variables;
- identify the degree of correlation between variables;
- describe a linear relationship with an equation of a straight line;
- analyze representations of data;
- draw conclusions based on given data sets and representations of data;*
- become aware of the questions that should be asked when analyzing data sets and representations of data.

** This goal is assessed in other sections of the unit.*

Pacing

- approximately three 45-minute class sessions

Vocabulary

- equation of a straight line
- least squares linear regression line
- median fit line
- slope
- *y*-intercept

About the Mathematics

A linear pattern in a scatter plot can be summarized by a line through the data. Different lines may be used including an "eyeball" line drawn freehand to capture the trend, a least squares linear regression line based on the mean of the data, and a median fit line based on finding medians as representative points.

Students are expected to be able to draw a line by eye and to find the equation of this line, as they learned in *Graphing Equations,* by finding the slope and the *y*-intercept.

The accuracy of the predictions made using a least squares linear regression line depends on how the data are scattered around the line. When the data are very close to the line, you are more confident about the predictions based on that line. Some relationships between variables cannot be summarized with straight lines. Growth, for example, is often associated with curves.

Materials

- Student Activity Sheets 6–9, pages 148–151 of the Teacher Guide (one of each per group of students)
- Student Activity Sheets 13 and 14, pages 155 and 156 of the Teacher Guide (one of each per student)
- drawing paper, page 107 of the Teacher Guide (two sheets per student)
- overhead projector, pages 117 and 119 of the Teacher Guide, optional (one per class)
- graph paper, pages 107, 109, 121, and 123 of the Teacher Guide (two sheets per student)
- rulers, pages 109 and 123 of the Teacher Guide (one per student)
- graphing calculators, pages 113, 115, 119, 121, 123, and 125 of the Teacher Guide (one per student)
- transparency of the graph on Student Book page 51, page 117 of the Teacher Guide, optional (one per class)
- blank transparencies, pages 117 and 119 of the Teacher Guide, optional (several per class)

Planning Instruction

You might inform students that they will be studying the correlations between the lengths and widths of various bird eggs. You may want to ask students, *What general statements can be made about the size and shape of bird eggs?* [They all are generally the same shape but vary in size.]

Students may work on problem 1 as a whole class. They may do problems 2–9 in small groups. The remaining problems may be completed in small groups or individually.

Problem 10, problems 16–20, and 23 are optional. If time is a concern, you may omit these problems or assign them as homework.

Homework

Problems 8 (page 110 of the Teacher Guide), 13 (page 116 of the Teacher Guide), and 21–23 (page 124 of the Teacher Guide) may be assigned as homework. The Extensions (pages 109, 111, 115, 117, and 127 of the Teacher Guide) and the Writing Opportunities (pages 119 and 127 of the Teacher Guide) may also be assigned as homework. After students complete Section F, you may assign appropriate activities from the Try This! section, located on pages 58–62 of the *Insights into Data* Student Book. The Try This! activities reinforce the key mathematical concepts introduced in this section.

Planning Assessment

- Problem 13 can be used to informally assess students' ability to describe the relationship between two variables and to analyze representations of data.
- Problem 14 can be used to informally assess students' ability to describe a linear relationship with an equation of a straight line.
- Problems 21 and 22 can be used to informally assess students' ability to represent data graphically and to describe data numerically. These problems may also be used to informally assess students' ability to describe the relationship between two variables and to describe a linear relationship with an equation of a straight line.
- Problem 24 can be used to informally assess students' ability to describe the relationship between two variables. It also assesses their ability to identify the degree of correlation between variables and to become aware of the questions that should be asked when analyzing data sets and representations of data.
- Problem 26 can be used to informally assess students' ability to identify the degree of correlation between variables.

F. LINES THAT SUMMARIZE DATA

Egg Hunt

Marisol noticed that a bird had recently laid eggs in a nest outside her window. The eggs appeared small, and she wondered whether they had the same shape as chicken eggs.

1. a. Do you think all bird eggs have the same shape?

b. How can you check your answer?

Throughout this section, you will use scatter plots to study the correlation between the lengths and widths of various bird eggs.

Solutions and Samples
of student work

1. **a.** Answers will vary depending on the students' knowledge of bird eggs. Most eggs have the same shape. In the list of 96 birds from Britain on Student Activity Sheets 6–8, for example, only the egg of the common sandpiper is nearly round.

   b. Answers will vary. Sample responses:

   Measure the lengths and widths of a variety of bird eggs, real or in photos.

   Refer to tables in an encyclopedia or to books about birds.

Hints and Comments

Overview A new context about the shape of eggs is introduced, and students think about the question "Do all bird eggs have the same shape?"

About the Mathematics In this section, students will use scatter plots to study correlation and use lines of best fit to predict the shapes of eggs.

Planning You may want to divide students into small groups and discuss problem **1** as a whole class.

Comments about the Problems

1. **b.** Trying to find differences in lengths and widths for bird eggs from a table is often difficult. In this section, students will learn that a scatter plot is useful in determining such differences.

On **Student Activity Sheets 6, 7,** and **8,** you will find tables with data on 96 birds of Britain. Since there are many birds, they are classified into families. Among the data, you will find the average lengths and widths of the eggs of each bird.

2. a. Draw an eider's egg using the dimensions found in the table.

b. Draw a lesser whitethroat's egg.

c. How are the eggs different? What might account for the differences?

You have discovered that eggs are different sizes. But what about their shapes? Is there a relationship between the lengths and widths of bird eggs? To answer these questions, you will begin by looking at the eggs of one family, the warbler.

3. a. Find all the warblers in the table and make a scatter plot that shows the relationship between the lengths and widths of the warbler eggs (length, width).

b. Describe the relationship between the lengths and widths of the warbler eggs. Is there any correlation?

c. What do you expect the width of a warbler egg will be if its length is 15 millimeters? How did you find your answer?

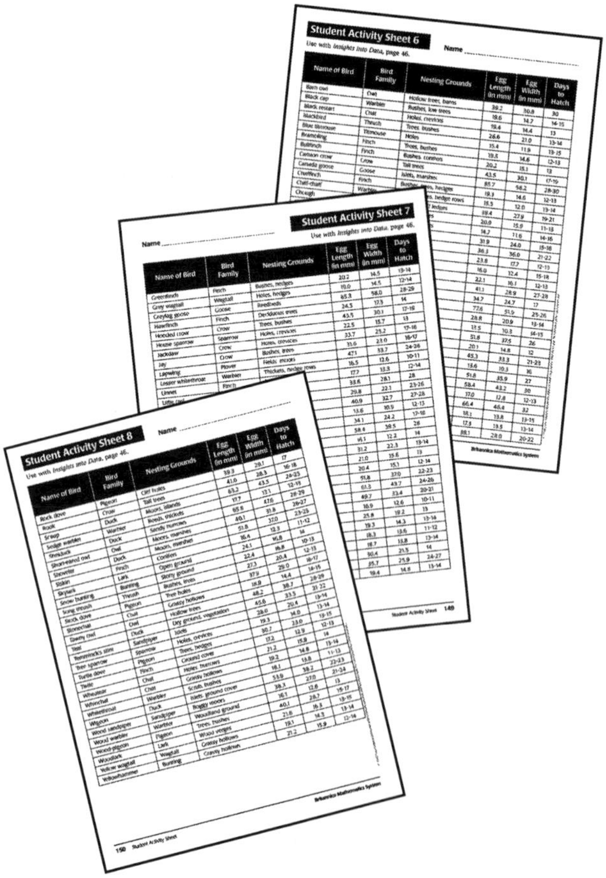

Solutions and Samples
of student work

2. a. Eider egg: length 77.6 millimeters; width 51.9 millimeters

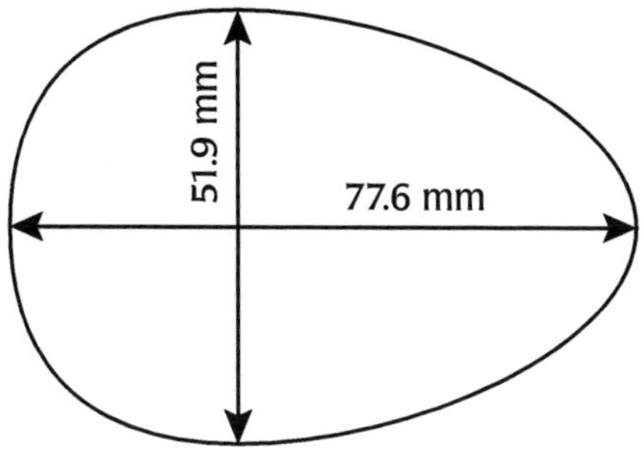

b. Lesser whitethroat egg: length 16.5 millimeters; width 12.6 millimeters

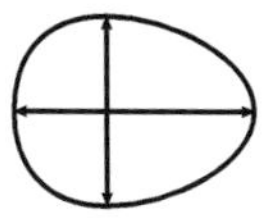

c. Although the eggs are about the same shape, the eider egg is bigger. The eggs are from different bird families.

3. a.

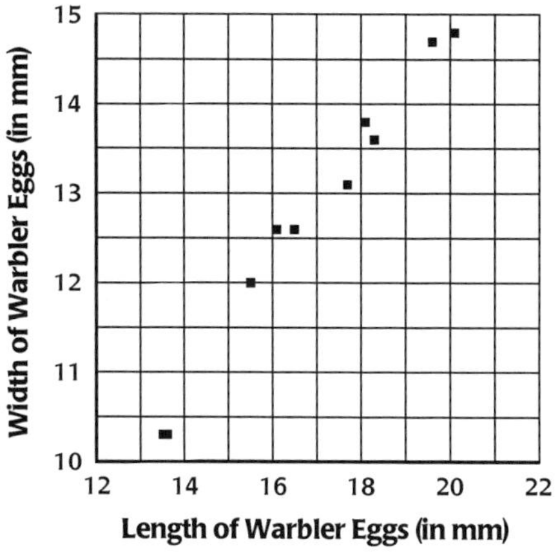

b. The points seem to form a straight line. There is a strong, positive correlation. The wider the egg, the longer it is.

c. The expected width is approximately 11.5 millimeters.

Strategies will vary. Sample student response:

I sketched the line made by the points in the scatter plot. Then I drew a vertical line from a length of 15 millimeters to the line in the scatter plot. At the point of intersection, I drew a horizontal line to the vertical axis and that would be about the width of the egg.

Hints and Comments

Materials Student Activity Sheets 6–8 (one of each per group of students); drawing paper (two sheets per student); Student Activity Sheet 13, optional (one per student); graph paper (one sheet per student)

Overview Students draw eggs, using dimensions from a table. Then they interpret a scatter plot and describe the relationship between the lengths and widths of eggs.

About the Mathematics From looking at the data in the table, it is hard to tell whether there is a relationship between widths and lengths of bird eggs. A scatter plot is helpful, because it makes the relationship visual.

There are two possibilities for making the scatter plot; length can be placed on the horizontal or vertical axis. The notation (x, y) indicates that the first variable (or value) is placed on the horizontal axis. The variable placed on the vertical axis is called the dependent variable.

As explained in the previous section, a correlation is the measure for how strong the relationship is. *Note:* Student Activity Sheet 13 gives directions for using a graphing calculator to make scatter plots.

Planning Students may work on problems **2** and **3** in small groups.

Comments about the Problems

2. Students should draw a standard egg shape and apply the dimensions given in the table.

3. a. You may need to explain the notation (length, width). This indicates that one should plot the length on the horizontal axis (x-axis) and the width on the vertical axis (y-axis). Some students may draw scatter plots that begin at (12, 10) since the warbler eggs all have lengths greater than 12 millimeters and widths greater than 10 millimeters.

Students need to recognize that they should scale the axes so that the data for the warbler eggs can be plotted.

You may want to remind students that many plots of real data do not begin at the origin because there are no data there and a large area of the plot would be empty. Students could think of such plots as a selected view of one portion of the graph.

The points in the scatter plot lie almost on a straight line. You might have used this information to find an answer to problem **3c.** It is possible to summarize the pattern, or relationship, by drawing a line that seems to best fit these points. This line will probably not go through all the points, but the points should lie close to the line.

4. a. On **Student Activity Sheet 9,** draw a straight line that seems to "fit" these points. Use your line to predict the width of a warbler egg that is 15 millimeters long.

b. Does your line give you the same answer that you predicted in problem **3c?**

5. a. Use your line to determine what happens to the width of an egg if the length increases by 2 millimeters. Explain how you did this. Include a drawing with your answer.

b. Copy and complete the following statements:

If the length of an egg increases by 2 millimeters, the width _______ by _______.

If the length of an egg increases by 1 millimeter, the width _______ by _______.

Solutions and Samples
of student work

4. a. Graphs may vary. See the sample graph in the solution for problem **5** below. The egg would be about 11.5 millimeters wide.

b. Answers will vary depending upon students' answers to problem **3c.**

5. a. If the length increases by 2 millimeters, the width increases by 1.5 millimeters. One strategy is shown in the graph below:

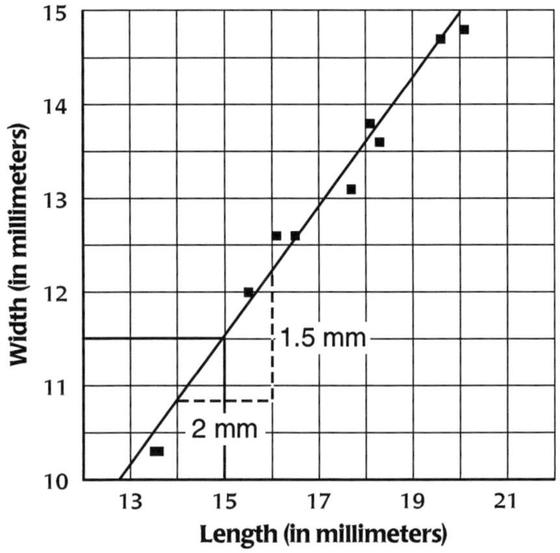

Select a length (14 millimeters is used in this example). Draw a horizontal line from the line that summarizes the data two units to the right (representing the increase of 2 millimeters). From this point, draw a vertical line to the line that summarizes the data. This line represents the increase in the width of the egg.

b. If the length of an egg increases by 2 millimeters, the width increases by 1.5 millimeters.

If the length of an egg increases by 1 millimeter, the width increases by 0.75 millimeter.

Hints and Comments

Materials Student Activity Sheet 9 (one per student); rulers (one per student); graph paper (one sheet per student)

Overview Students draw a straight line in the scatter plot and use this line to investigate the relationship between length and width.

About the Mathematics Sketching a line that fits the data is a skill that students should develop in this unit. They should recognize that you can use several strategies to draw a best-fit line. There are many possible definitions for a line of best fit. These include a line that goes through as many points as possible or a line that has as many points above the line as below the line.

Planning Students may work on problems **4** and **5** in small groups. After problem **5,** you might have a class discussion about how to draw a line that fits best and how to use this line to describe the relationship between the variables. The Extension may be used as an informal assessment.

Comments about the Problems

4. a. Students may place a ruler over the scatter plot and move it until the edge seems to cut through the middle of the cloud of points. The line does not need to pass through any data points.

5. a. *Note:* The scales of the axes are different. This causes the horizontal increase of the length of 2 mm to seem smaller than the vertical increase of the width of 1.5 mm.

b. It is not necessary to have students write an equation for their line.

Extension This Extension activity assesses students' ability to represent data graphically and to describe a linear relationship with an equation of a straight line. Students may use the data from Student Activity Sheets 6–8 to make a scatter plot of the lengths and widths of finch eggs. They could draw a line that summarizes the data and use it to predict the width of a finch egg if the length is 17 millimeters.

In problem **4,** you looked at a scatter plot and drew a line that summarized the relationship of the data points. Your line may have been very different from the lines drawn by your classmates.

6. How can you decide whether a line drawn to summarize the data in a scatter plot is a good line? What criteria would be helpful in making your decision?

7. a. Use your criteria to decide whether or not you drew a good line.

b. How does your line compare with the line drawn by a classmate?

There are many ways to find a good line. Different criteria are used for different purposes. One criterion is how well your line predicts a value for something you already know about.

8. a. Look at the graph of your line from problem **4.** What does it predict for the width of a lesser whitethroat warbler egg that is 16.5 millimeters long?

b. Now look at the data on **Student Activity Sheets 6–8.** What is the actual value for the width of the lesser whitethroat warbler egg? How well did your line predict this value?

c. Overall, does the line you drew seem to predict values well when you compare it with the actual data? Explain.

6. Answers will vary. Sample response:

The line should have a slope that is in the same direction as the cloud of points and should summarize the points without necessarily going through each point in the plot.

7. a–b. Answers will vary. Some students may feel it is hard to tell whether their line is a good line. Other students may measure how far off each point is from the line. Still others may count the number of data points on the line.

8. a. Answers will vary. Students should find the point on the line where the length is 16.5 millimeters and estimate the width of the egg at that point on the line. For the line drawn in problem **4,** the width would be about 12.6 millimeters.

b. The actual width is 12.6 millimeters. The difference will vary, depending on the location of students' lines.

c. Answers will vary, but generally the line is a good predictor of egg width (as confirmed by checking the length/width measurements of other birds' eggs).

Materials Student Activity Sheets 6–9 (one of each per group of students)

Overview Students continue with the investigation of the scatter plot and the line of best fit that describes the data.

About the Mathematics As discussed on page 109 of the Teacher Guide, there are many possible criteria for deciding what a good line of best fit is. Thinking of possible criteria is more important than formalizing them. When a line is chosen, it can be used to make predictions.

Planning Students may work in small groups on problems **6–8.** Problem **8** may be assigned as homework.

Comments about the Problems

6. Encourage students to share their criteria for a good line. Students might think that a good line connects all data points. This could result in a zigzag pattern, which is not a line. Some students might connect the first and last points in the data set. However, the line through these points may not summarize the data.

7. a–b. Encourage students to discuss the criteria and how the criteria would be applied for different lines.

8. a–b. Homework This problem can be assigned as homework.

Extension You may want students to think of a measurement to describe how well a line fits the data. Encourage students to be creative with their answers.

In problem **4,** you used a best-fit line to predict the width of the warbler egg. Statisticians often use a line called the *least squares linear regression line* to predict data. You can find the equation for this line by using a graphing calculator or computer software.

9. a. Use **Student Activity Sheet 14** to find the equation of the least squares linear regression line for the warbler eggs. Make a scatter plot on your calculator and graph the equation you found.

b. Now use your new line to find the width of the lesser whitethroat warbler egg that is 16.5 millimeters long. How close is the prediction to the width given in the table?

c. Overall, how well does the least squares linear regression line seem to predict the widths of the warbler eggs?

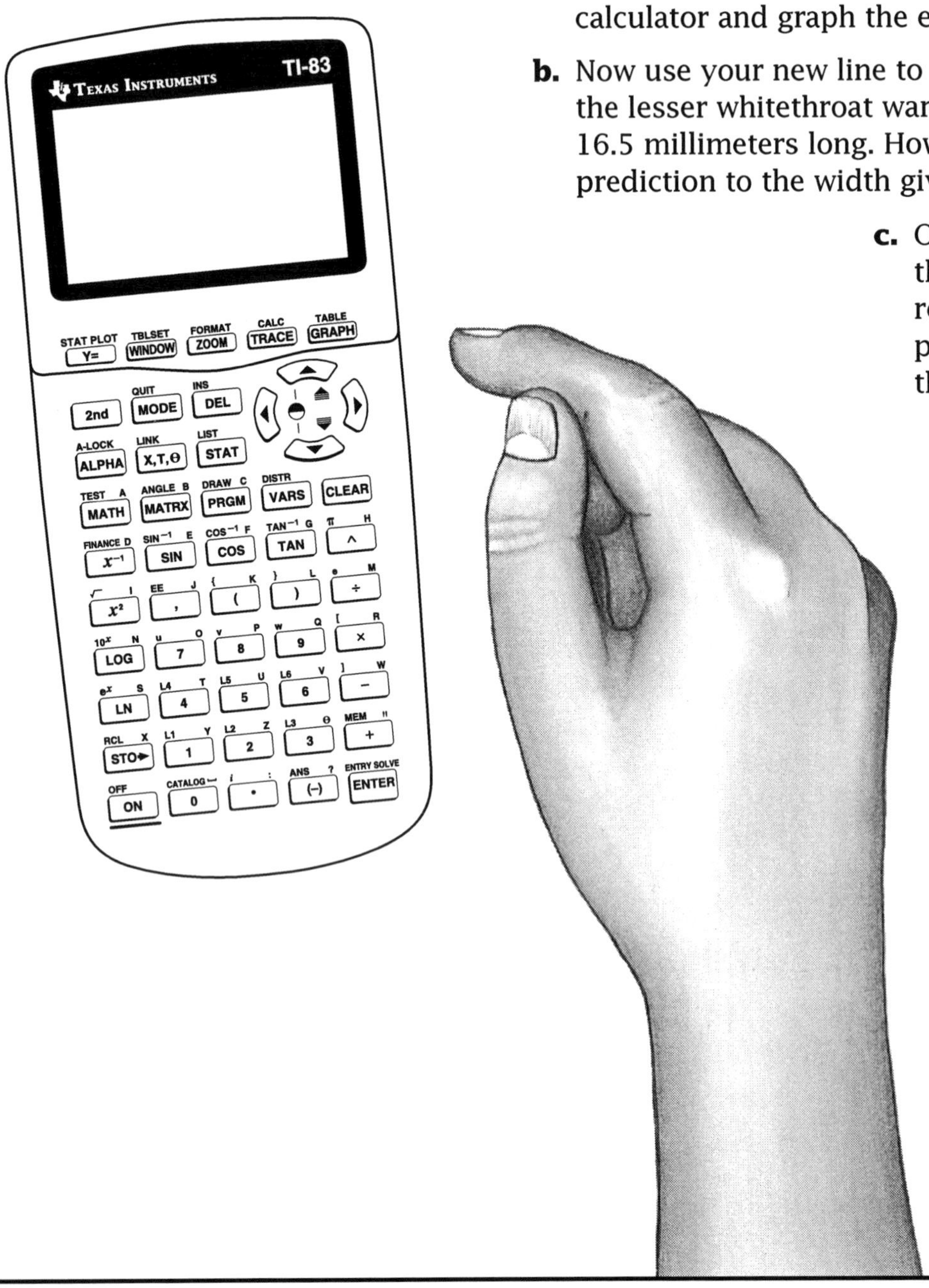

9. a. The equation of the least squares linear regression line for warbler eggs is
$y = 0.693x + 1.09$.

b. The line shows that the width is 12.525 millimeters.
Replace x with 16.5.
$y = 0.693(16.5) + 1.09$
$y = 11.435 + 1.09$
$y = 12.525$

The line predicts the width to be 12.5 millimeters. This is only 0.1 millimeter less than the width given in the table. For this length, the line seems to predict the width quite well.

c. For most of the eggs, the predictions made from the line seem quite accurate.

Hints and Comments

Materials graphing calculators (one per student); Student Activity Sheet 9 (one of each per group of students); Student Activity Sheets 13 and 14 (one of each per student)

Overview Students use a graphing calculator to make the least squares linear regression line for the warbler eggs.

About the Mathematics Graphing calculators and statistical software usually have the capability of using regression in scatter plots, so you can have the program draw a line or polynomial that best fits a set of data in a scatter plot. Note that such a line can *always* be drawn, even if there is no relationship between the variables. Therefore, it is important to have students first decide whether a relationship will exist and what kind of relationship it will be before having them draw lines with the graphing calculator or computer.

Planning You may have students work on problem **9** in small groups and then compare the results of problems **8** and **9** in a class discussion.

Note: Student Activity Sheets 13 and 14 have directions for using a graphing calculator to make a scatter plot and to graph a line.

Comments about the Problems

9. a. If students do not have a graphing calculator, you may give them the equation of the line, which is as follows:

$y = 0.7x + 1.1$, or
width = $0.7 \times$ length + 1.1.

The line can be drawn in the same plot used in problems **4–8.**

b. Encourage students to use at least three digits in their calculations with the graphing calculator and to round the final result to the nearest tenth to fit the accuracy of the data. If students do not have graphing calculators, they can use either the equation given above or the graph to find an estimate for the width. Accuracy will, of course, be affected by the scale used on the axes of the scatter plot.

Another line that statisticians use is called the *median fit line.* This is a good line to use when there are outliers in the data.

10. **a.** Following the directions on the bottom of **Student Activity Sheet 14,** use your graphing calculator to find the equation of the median fit line for the warbler eggs. Graph the line in your scatter plot.

b. How does the median fit line compare with the least squares linear regression line?

c. Does the median fit line seem to do a good job of predicting the widths of the eggs?

11. **a.** Choose either the median fit line or the least squares linear regression line and find the *slope* of the line. How can you find this number in the equation?

b. How can you find this number from the graph?

c. What does the slope tell you about the relationship between the length and the width of warbler eggs?

12. **a.** What does the other number in the equation you selected in problem **11a** represent on the graph (the number that is not the slope)?

b. What does the number mean in terms of egg length and width?

Killdeer's eggs

Mallard duck's eggs

Red-winged blackbird's eggs

10. a. The equation of the median fit line is $y = 0.703x + 0.903$.

b. The median fit line is almost the same as the least squares linear regression line. If students use the ZOOM function on their graphing calculators, they may be able to see that the median fit line is slightly lower.

c. Answers may vary. The line seems to give good predictions. The points do not seem to be very far from this line.

11. a. The slope of the least squares linear regression line is 0.693. The slope of the median fit line is 0.703. (The slope is the coefficient of x, or the number by which x is multiplied.)

b. Answers may vary. Sample response:

The slope of the line can be rounded to 0.7, or $\frac{7}{10}$, to simplify the work. From any point on the line, draw a path that is 10 mm horizontally and then 7 mm vertically as shown below. The new location should also be on the line that summarizes the data. This process can be repeated with any point on the line.

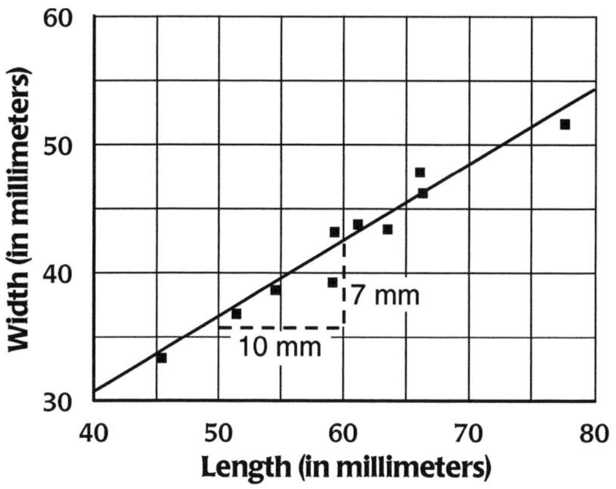

c. If the length of an egg increases by 10 mm, the width increases by 7 mm; or if the length increases by 1 mm, the width increases by 0.7 mm.

12. a. The other number is the y-intercept. It is the location on the vertical axis at which the line that summarizes the data crosses the vertical axis.

b. It gives the width of the egg when the length is 0. It is meaningless in this situation. If the data were about something else, the y-intercept might be a meaningful number.

Materials graphing calculators (one per student); Student Activity Sheets 9 and 14 (one of each per student)

Overview Students use a graphing calculator to draw lines that summarize the relationship between data. They interpret the slope and the y-intercept in terms of the data.

About the Mathematics The median fit line is introduced here as another line that summarizes data. More important than knowing that different lines exist is to know what the numbers in the equation of a line mean in terms of the data.

Planning Problem **10** is optional. If time is a concern, this problem may be omitted or assigned as homework. Discuss the solutions to problems **11** and **12** as a class before proceeding to the next page. The Extension may be used as informal assessment or assigned as homework. Students may work on problems **10–12** in small groups or individually.

Comments about the Problems

10. Students may find it easier to draw the lines by hand. You might want to give them the equation of the line and not go into detail about how regression lines and median fit lines are computed. Alternatively, you may want to omit problem **10** and not mention median fit lines at all.

11. Students should feel free to use either line. If students do not have a graphing calculator, or if problem **10** was skipped, you may decide to have them use the equation given earlier. (See Hints and Comments, problem **9a** on Teacher Guide page 113.)

b–c. Have students refer to their solution to problem **5a** (Student Book page 47).

Extension/Informal Assessment This Extension activity assesses students' ability to describe a linear relationship with an equation of a straight line.

You may want to continue the Extension activity proposed on page 109 of the Teacher Guide. Ask students to find an equation for the line best fitting these data and have them do the equivalent of problems **11** and **12** with the plot and the line(s) for the finch eggs.

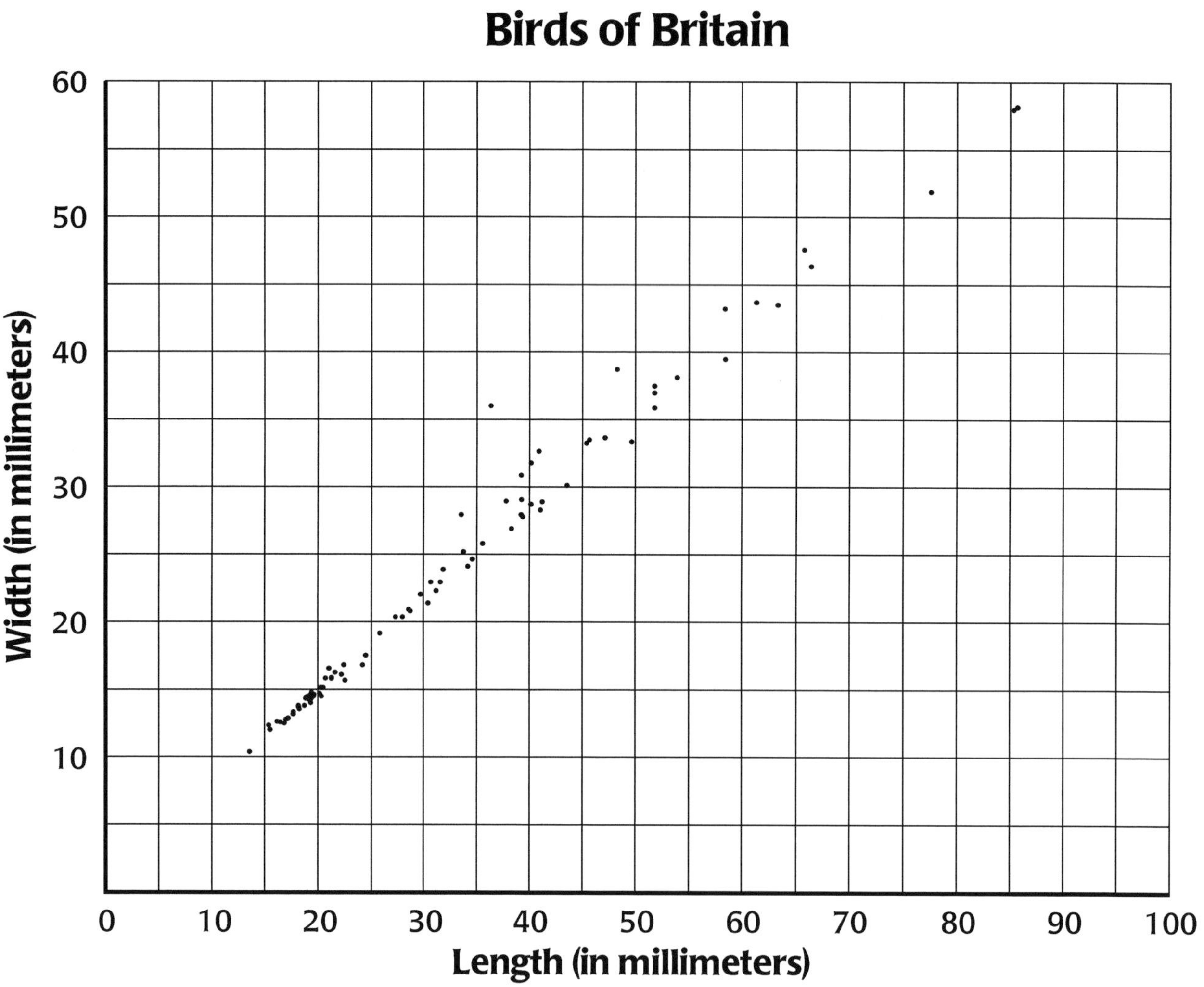

The scatter plot above includes the data for all 96 birds of Britain.

13. a. Locate the warbler eggs in the scatter plot.

b. In the scatter plot, what happens to the relationship between length and width as the bird eggs get larger?

13. a. The warbler eggs are in the group of eggs whose lengths range from 13 to 21 millimeters, as shown below:

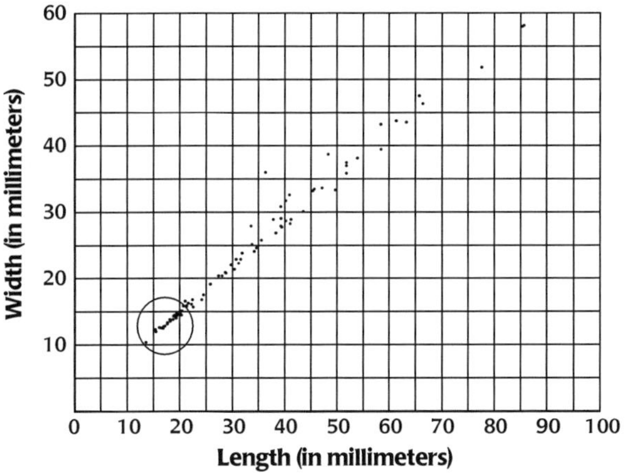

b. As the bird eggs get larger, the relationship is not as strong as the relationship between the lengths and widths of warbler eggs. The warbler eggs seem to fit a straight line more closely than do the dimensions of the larger eggs.

Materials transparency of the graph on Student Book page 51, optional (one per class); blank transparencies, optional (several per class); overhead projector, optional (one per class)

Overview Students find the general location of the warbler eggs on the scatter plot of all the bird eggs. Then they reason about the relationship as it appears in this scatter plot.

Planning Students may work in small groups or individually. Problem **13** may be used as informal assessment or assigned as homework.

Comments about the Problems

13. Informal Assessment This problem assesses students' ability to describe the relationship between two variables and to analyze representations of data. This problem may be assigned as homework.

a. You may want to make a transparency of the graph for class discussion. Blank transparencies can be placed on top of the Student Book for students to circle warbler egg points.

b. Ask students, *Which bird eggs do not fit the pattern of most bird eggs?* [The common sandpiper egg, for example, with an egg length of 36.3 millimeters and a width of 36.0 millimeters, is far from the straight line that is determined by the other points.]

Extension As a continuation of the Extension activities proposed earlier, you may have students locate the finch eggs in the scatter plot as well. You may also have students draw a best-fit line for the scatter plot of all the birds, find its equation, and formulate the relationship in terms of the following statement: If the length increases by ___ millimeters, the width increases by ___ millimeters.

The equation for the least squares linear regression line for the scatter plot on page 51 is:

$$\text{Width } (y) = 0.7 \times \text{Length } (x) + 1.6$$

14. **a.** Compare the line above to the least squares linear regression line for the warblers ($y = 0.6x + 1.0$) by graphing both on your calculator.

b. What can you tell about the two lines?

c. Why do you think the two lines are different?

d. If you make predictions based on the line for all eggs, what will happen to your predictions as the egg lengths increase?

15. What can you tell about the egg shapes of birds from different families?

Activity

Measure some chicken eggs at home. Record the lengths and widths. In class, collect everyone's data. Find the typical length and width for a chicken egg.

16. How does the typical chicken egg fit in the graph of the birds of Britain?

14. **a.** *Note:* The equation for the least squares linear regression line for the warblers is $y = 0.7x + 1.1$; the equation given on Student Book page 52 is in error.

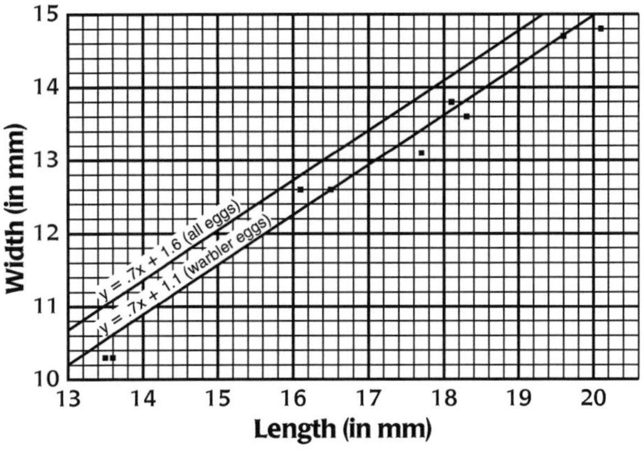

b. Answers may vary. Sample response:

The lines are almost parallel. Since the *y*-intercept for the line that summarizes the data for all 96 birds is 1.6, the line starts slightly higher on the vertical axis.

c. Answers will vary. Sample response:

The second line includes the data from all 96 birds of Britain. The first line includes only the data from one bird family, the warblers.

d. Answers will vary. Sample response:

As egg length increases, points are farther from the line, and there will be more error in predicting the width of an egg.

15. Answers will vary. Students may simply note that the sizes of eggs from different bird families may vary greatly.

16. Answers will vary, depending on class data and on what one considers "typical." Record the data of the lengths and widths of chicken eggs from each person in the class on a transparency or on the chalkboard. From these data, students can determine the measurements of a typical chicken egg.

Hints and Comments

Materials graphing calculators (one per student); blank transparency, optional (one per class); overhead projector, optional (one per class)

Overview Students compare the line for all bird eggs to the line that best fits the warbler eggs.

About the Mathematics If only part of the data is used to make a line fitting the data, this line will differ from the line that best fits all data points. In some cases, it may be more meaningful to use only part of the data. For instance, in the scatter plot from Section A, a pattern may appear for all states in the Midwest. In that case, these data points can be selected and a line can be fit through them.

Planning Problem **14** can be used as an informal assessment. Problem **16** is optional and may be omitted or assigned as homework if time is a concern. Students may work on problems **14–16** in small groups or individually.

Comments about the Problems

14. Informal Assessment This problem assesses students' ability to describe a linear relationship with an equation of a straight line.

a. The scale of this graph has been enlarged to clearly show both lines. If students do not use a graphing calculator, they may draw the lines from the given equations in the scatter plot.

Writing Opportunity You may want students to write a brief report in their notebooks as an answer to problem **16.**

Interdisciplinary Connection Students can ask their biology teacher about the breeding and nesting habits of certain birds and whether these habits affect egg size or shape.

Gone Fishing

Graphs of growth often result in curves, because living things do not grow at a constant rate throughout their lifetimes.

Researchers in fisheries are interested in studying the growth rates of fish. One study on bluegills compared the lengths of fish at the beginning of the year with their lengths at the end of the year. The table on the left contains the results of this research.

Bluegill Growth

Initial Length *x* (in mm)	Length after 1 Year *y* (in mm)
48	69
52	71
51	69
53	75
69	101
71	107
69	100
75	104
101	138
107	138
100	130
104	140
138	160
132	157
130	156
140	161
160	173
157	168
156	172
161	178
173	176
168	174
172	173
178	178

17. **a.** How much did the fish that was initially 161 millimeters grow in one year?

b. Using a graphing calculator, plot the growth from part **a** (initial length, length after one year).

17. a. The fish grew 17 millimeters, to a length of 178 millimeters.

b. See the scatter plot below.

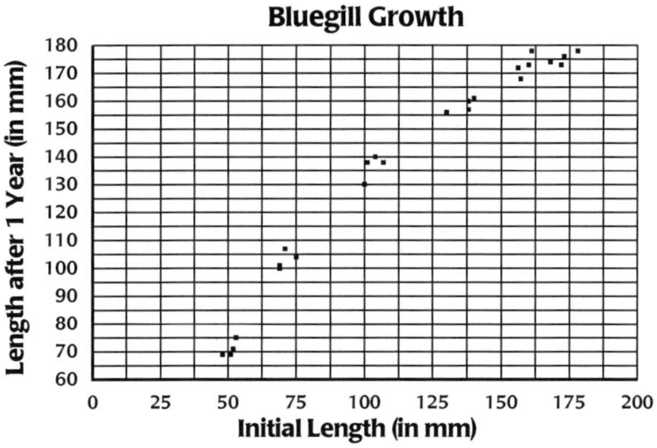

Materials graphing calculators (one per student); graph paper, optional (one per student)

Overview Students study data in which a nonlinear relationship between variables occurs.

About the Mathematics As previously stated, in each scatter plot a line can be drawn. Computer graphing programs will always draw a best-fit line if asked. Often this line does not really fit the data or is meaningless in the context. The example used here shows data for which a curve fits better than a line.

Planning Students may work on problem **17** in small groups or individually. Problem **17** is optional. If time is a concern, it may be omitted or assigned as homework.

Comments about the Problems

17. b. If graphing calculators are not available, students may draw the scatter plot.

18. **a.** What pattern do you see in the scatter plot from problem **17b?**

b. Graph a straight line to represent the data. Use the line to predict the length of a 110-millimeter fish after one year. What is the predicted length for a 140-millimeter fish after one year?

c. Is a straight line a good model to represent the data in the graph? Explain your reasoning.

19. **a.** Plot the data from the table on page 53 on graph paper and sketch a better model to represent the relationship between length and growth.

b. In your new model, what is happening to the growth of the fish as the length changes?

c. If a fish is 110 millimeters long, predict how long it will be in one year.

20. Describe the difference between using a straight line and using the model you created in problem **19** to predict growth.

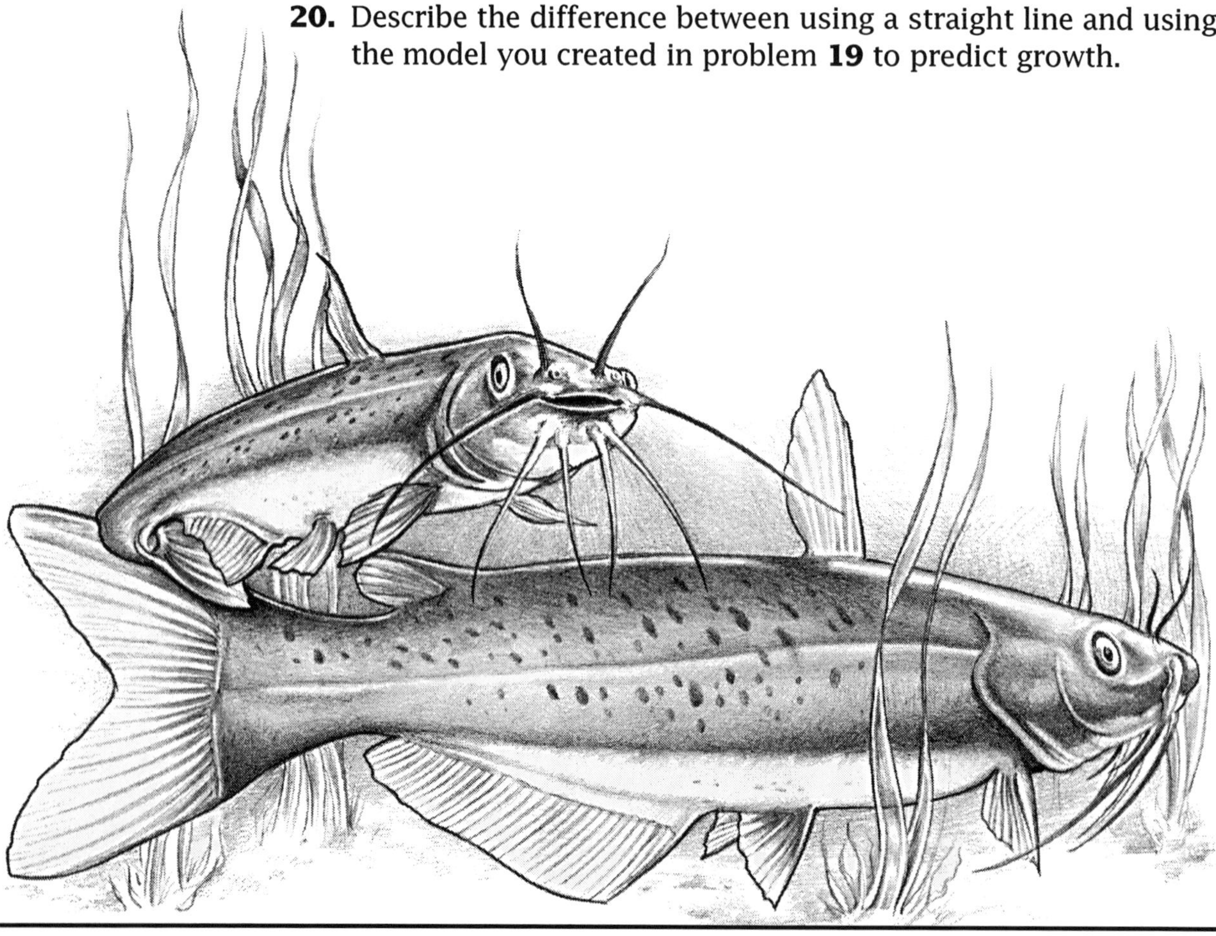

18. a. Answers will vary. Sample response:

Small fish grew faster than fish that were long initially. A curve may fit the pattern.

b. Answers will vary. Sample response:

Using the straight line model, as shown below, if a fish is initially 110 millimeters long, it will grow to about 132 millimeters. If a fish is initially 140 millimeters, it will grow to about 157 millimeters.

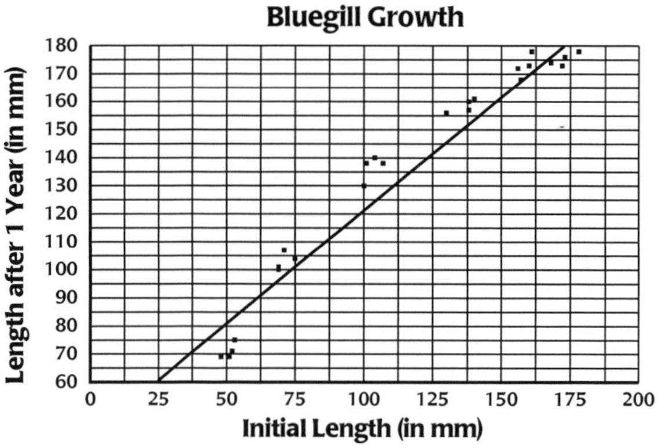

c. No. A straight line does not summarize the pattern in the data well. It is hard to find a line that fits.

19. a. Sketches may vary. Sample drawing:

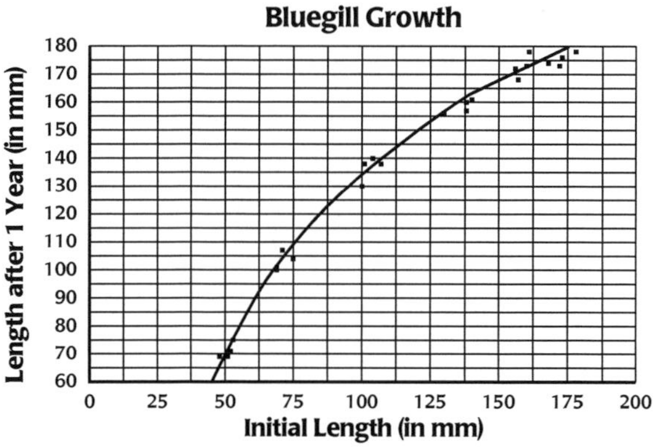

b. Answers will vary. Sample response:

As the fish get longer, the rate of growth slows down.

c. The fish should be about 140 millimeters at the end of one year.

20. The use of a straight line to summarize the data suggests that bluegills will always grow larger. A curve suggests that large fish grow less and eventually not at all.

Hints and Comments

Materials graphing calculators (one per student); graph paper, optional (one sheet per student); rulers, optional (one per student)

Overview Students decide whether a line or some other model best describes the data on the growth of the fish.

About the Mathematics In cases like this, where "real-world" data are used, it is very important to find a contextual reason why a curve fits better than a straight line. In this case, the model needs to show that there is a limit to how long bluegills can grow. A curve that gets flatter for larger initial lengths shows this limit, whereas a straight line does not.

Planning Students may work on problems **18–20** in small groups or individually. These problems are optional. If time is a concern, they may be omitted or assigned as homework. When students have completed these problems, you may want to summarize the most important results of problems **17–20** in a class discussion.

Comments about the Problems

18. b. If students are using calculators, they should graph either the least squares linear regression line or the median fit line. If students drew the scatter plot, they should approximate a line of best fit.

c. Students should realize that the line will be a poor predictor for points that are far away from the line. Students should also notice that with the straight line, the data points in the middle are above the line and data points at the ends are below the line.

19. a. Graphing large scatter plots may be too time consuming, since it is not a goal in this section. When fitting lines or curves in a scatter plot, it is important to have a contextually sound explanation for the kind of curve or line that fits. See the About the Mathematics section above.

Comparing Scores

Do test scores for mathematics and science relate to each other? Below are some results from an international comparison of math and science.

Average International Test Scores for 8th Grade

Comparing Students in the United States with Foreign Students

Overall Math Scores (in percent of correct answers)		Overall Science Scores (in percent of correct answers)	
Singapore	79	Singapore	70
Japan	73	Korea	66
Korea	72	Japan	65
Hong Kong	70	Czech Republic	64
Belgium	66	England	61
Czech Republic	66	Hungary	61
Slovak Republic	62	Belgium	60
Switzerland	62	Slovak Republic	59
Hungary	62	Sweden	59
France	61	Canada	59
Russian Federation	60	Ireland	58
Canada	59	United States	58
Ireland	59	Russian Federation	58
Sweden	56	New Zealand	58
New Zealand	54	Norway	58
Norway	54	Hong Kong	58
England	53	Switzerland	56
United States	53	Spain	56
Spain	51	France	54

Source: Data from Science/Mathematics Achievement in the Middle School Years: IEA's Third International Mathematics and Science Study (Boston College: Center for the Study of Testing, Evaluation, and Educational Policy, 1996), 41.

21. a. Using the results for each country in the table above, make a scatter plot (math scores, science scores). Use a graphing calculator if one is available.

b. In which subject are the scores better? Explain how you can see this in the graph.

22. a. Graph the line that describes the relationship between the scores. Use it to predict the science score when the math score is 50%.

b. What is the slope of your line? What does it tell you about the scores?

23. List the countries in order, from highest to lowest, according to their scores on math and science combined. Describe how you ordered the countries.

Solutions and Samples
of student work

21. a.

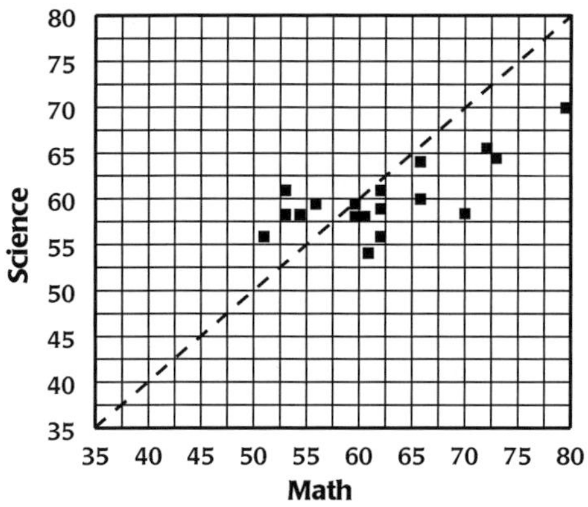

b. The math scores are better. When you draw a line along the diagonal where the math scores equal the science scores, most points are below this line.

22. a. The science score would be around 55%. See graph below.

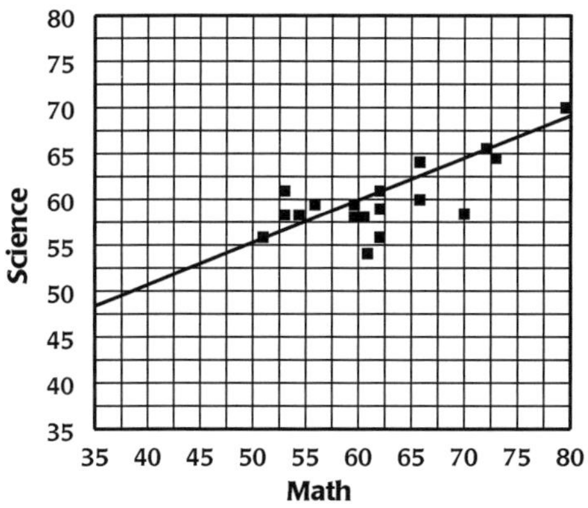

b. The slope is about 0.5. If the math score increases by 1, the science score increases by 0.5.

23. Students should add the math and science scores for each country and then list the countries in order according to their sums.

1. Singapore
2. Korea and Japan (tie)
3. Czech Republic
4. Hong Kong
5. Belgium
6. Hungary
7. Slovak Republic
8. Canada, Russian Federation, and Switzerland (tie)
9. Ireland
10. Sweden and France (tie)
11. England
12. New Zealand and Norway (tie)
13. United States
14. Spain

Hints and Comments

Materials graphing calculator (one per student)

Overview Students make and interpret a scatter plot. They use a line to describe the relationship between the variables.

About the Mathematics Drawing lines can be done by eye. When data are entered in the graphing calculator or the computer, the calculator or computer can draw the best-fitting line.

Planning Students may work on problems **21–23** in small groups or individually. Problems **21** and **22** may be used as informal assessments or assigned as homework. Problem **23** is optional. If time is a concern, the problem may be omitted or assigned as homework.

Comments about the Problems

21–22. Informal Assessment These problems assess students' ability to represent data using a scatter plot and to describe data numerically. These problems also assess their ability to describe the relationship between two variables and to describe a linear relationship with an equation of a straight line. These problems may also be assigned as homework.

21. a. Before making a scatter plot, students should rewrite the information as a pair of scores for each country. For example, Singapore (79, 70), Japan (73, 65). . . . Students should not enter the data into their calculators in the order given in the table.

22. b. Instead of 1 to 0.5, one can also say 10 to 5, or any other pair of numbers with the same ratio.

Students can formulate the meaning of the slope in terms of the data, as they did with the eggs. If necessary, have them look back to their answers to problem **5** on Student Book page 47 and problem **11** on Student Book page 50.

Note: You may want to explain to students how to find the slope in the graph without using a graphing calculator.

Scattered INFORMATION

24. Describe the relationship between the variables in each of the following scatter plots. (*Note:* These data are not based on facts.)

a. **1993 NBA Lakers Games (until March 11)**

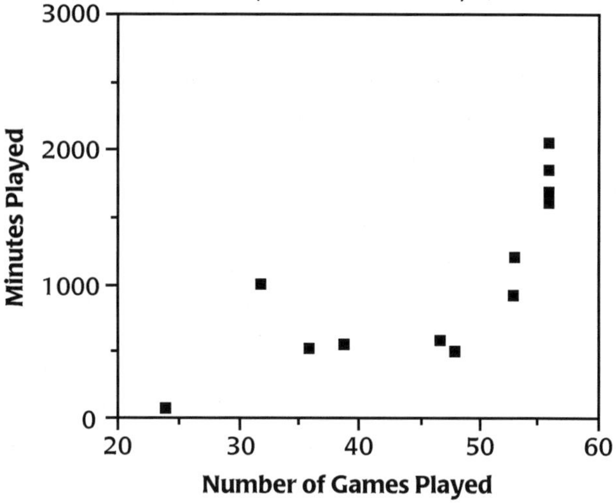

b. **Vehicle Fuel Economy**

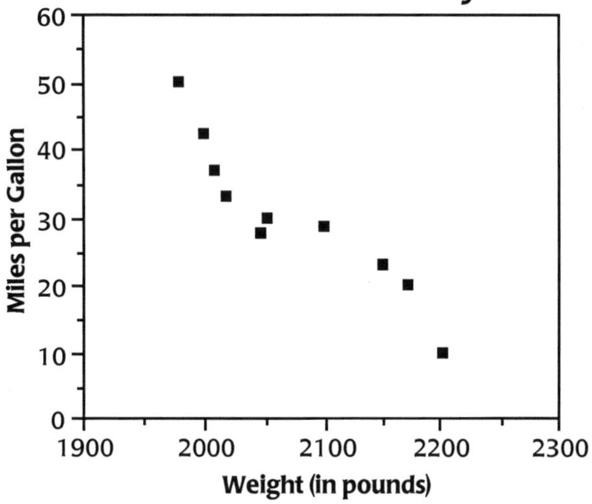

c. **Birds of Britain**

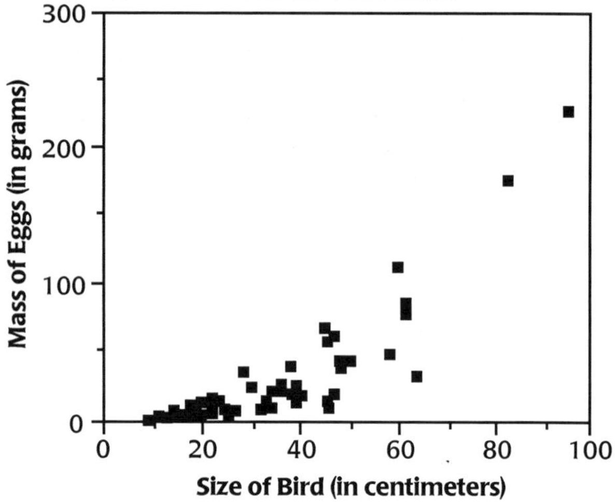

d. **Birds of Britain**

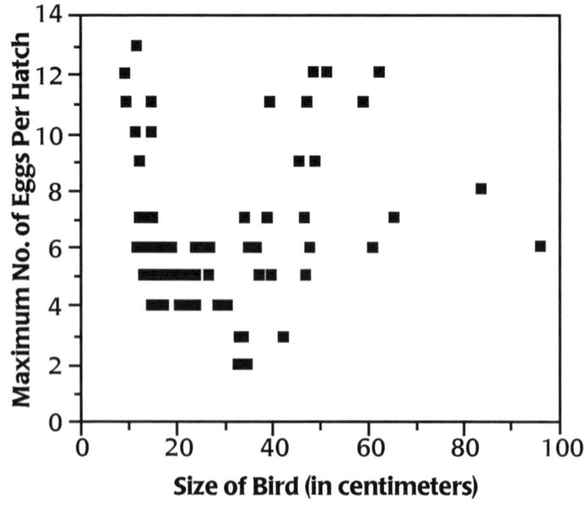

e. **Shoe Size and Head Circumference**

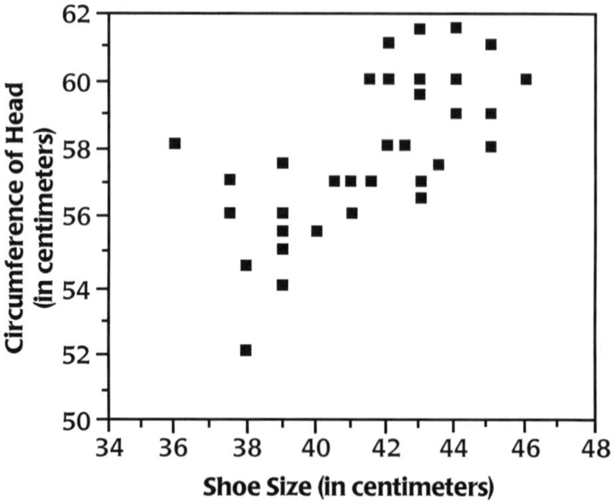

24. Answers will vary. Sample responses:

a. The more games in which a player played, the more minutes he played. This shows a positive correlation, although not a very strong one.

b. As the weight of the vehicle increases, the fuel efficiency decreases. This shows a rather strong negative correlation, suggesting a linear relationship.

c. Small birds lay light eggs; as the size of the bird increases, the egg becomes heavier. This shows a positive correlation, which seems best described by a curve.

d. There seems to be no relationship (correlation) between the size of a bird and the number of eggs hatched.

e. There appears to be a weak, positive correlation indicating that a person with a large shoe size will have a large head circumference. However, the data seem to form two clusters. More data should be collected before making a claim that shoe size is an indicator of head circumference.

Overview Students describe the relationship between variables in different scatter plots.

Planning Students may have difficulty making up stories for given plots of data. For this problem, you might lead a whole-class discussion for students to share and discuss their stories.

Students may work on problem **24** in small groups or individually. If time is a concern, you may want to assign one scatter plot to each group. This problem and the Extension may be used as informal assessments.

Comments about the Problems

24. Informal Assessment This problem assesses students' ability to describe the relationship between two variables; to identify the degree of correlation between variables; and to become aware of the questions that should be asked when analyzing data sets and representations of data.

Extension Students may collect their own data for a relationship they think will have a correlation. They could then make a scatter plot of the data and draw a line to summarize the relationship. This Extension activity assesses students' ability to analyze and interpret scatter plots.

Writing Opportunity Students could write a paragraph in their notebooks describing their solutions for the Extension activity.

Summary

If the relationship between two variables appears to be linear, a line can be found to describe the relationship.

Straight lines can be used to predict unknown values, check existing values, and compare different values.

The slope of the line can be expressed in terms of the data as you saw on page 47. This can lead to statements such as the following:

When the ______ increases by ______, the ______ ______ by ______.

The best fit line drawn by hand is used to capture the trend of the data.
The least squares linear regression line is based on the mean of the data.
The median fit line is used to find medians as representative points.
A curve is used to describe data that is not constant, such as growth.

Summary Questions

25. Explain how you can find the slope of a line from a statement like the one bolded in the summary above.

26. Match each of the descriptions at the bottom of the page to a corresponding scatter plot.

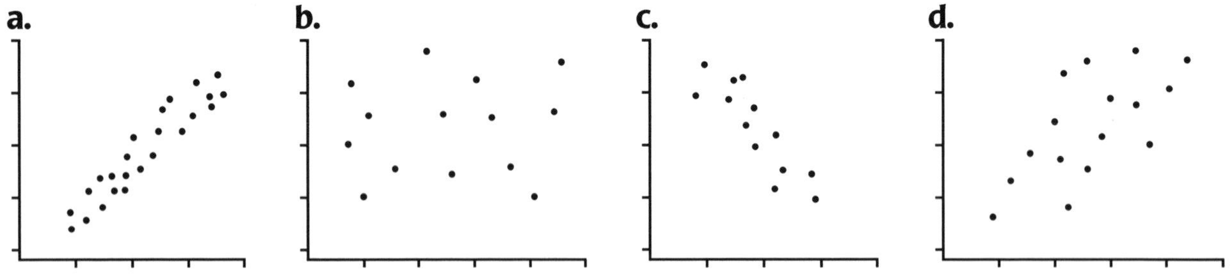

i. linear, negative, and moderate

ii. linear, positive, and strong

iii. no apparent correlations

iv. linear, positive, and weak

25. Answers may vary. Students should note that the slope of a line can be found by dividing the vertical increase (or decrease) by the corresponding horizontal increase.

26. a. scatter plot matches with statement ii

b. scatter plot matches with statement iii

c. scatter plot matches with statement i

d. scatter plot matches with statement iv

Overview Students read the Summary and reflect on correlations and the use of best-fit lines in scatter plots.

Planning Students may work on problems **25** and **26** in small groups or individually. Problem **26** may be used as informal assessment. After students complete Section F, you may assign appropriate activities from the Try This! section, located on pages 58–62 of the *Insights into Data* Student Book, for homework.

Comments about the Problems

26. Informal Assessment This problem assesses students' ability to identify the degree of correlation between variables.

Assessment Overview

Students work on three assessment activities that you can use to collect additional information about what each student knows about representing data using scatter plots, describing the relationships between variables, and designing, conducting, and analyzing ways of gathering data.

Goals	Assessment Opportunities
• represent data graphically	Hats, Manatees and Power Boats, Design-a-Test
• describe data numerically	Hats, Manatees and Power Boats, Design-a-Test
• describe the relationship between two variables	Hats, Manatees and Power Boats, Design-a-Test
• identify the degree of correlation between variables	Hats, Manatees and Power Boats, Design-a-Test
• describe a linear relationship with an equation of a straight line	Design-a-Test
• design, conduct, and analyze ways of gathering data: surveys, simulations, experiments	Hats, Manatees and Power Boats, Design-a-Test
• use random samples in gathering data	Design-a-Test
• analyze representations of data	Design-a-Test
• draw conclusions based on given data sets and representations of data	Hats, Manatees and Power Boats, Design-a-Test
• recognize possible bias in sample surveys	Design-a-Test
• determine whether representations of data (numerical and visual) are appropriate	Hats, Manatees and Power Boats, Design-a-Test
• become aware of the questions that should be asked when analyzing data sets and representations of data	Hats, Manatees and Power Boats, Design-a-Test

Pacing

- When combined, the three assessment activities will take approximately three 45-minute class sessions. See the Planning Assessment section for further suggestions as to how you might use the assessment activities.

About the Mathematics

The three assessment activities evaluate the major goals of the *Insights into Data* unit. Refer to the Goals and Assessment Opportunities section on the previous page for information regarding the specific goals assessed in each assessment activity. Students may use different strategies to solve each problem. Their choice of strategies may indicate their level of comprehension of the problem. Consider how well students' strategies address the problem, as well as how successful students are at applying their strategies in the problem-solving process.

Materials

- Assessments, pages 157–160 of the Teacher Guide (one of each per student)
- Student Activity Sheets 12 and 13, pages 154 and 155 of the Teacher Guide, optional (one of each per student)
- graph paper, pages 133, 135, and 137 of the Teacher Guide, optional (several sheets per student)
- graphing calculators, pages 133, 135, and 137 of the Teacher Guide, optional (one per student)
- rulers, pages 133, 135, and 137 of the Teacher Guide, optional (one per student)

Planning Assessment

You may want students to work on these assessments individually so that you can evaluate each student's understanding and abilities. Make sure that you allow enough time for students to complete the assessment activities. Students are free to solve each problem in their own way. They may choose to use any of the models introduced and developed in this unit to solve problems that do not call for a specific model.

The Hats assessment and Manatees and Power Boats assessment evaluate only a selection of the goals and mathematical content of this unit. They focus on the mathematical concepts covered in Sections A, D, E, and F. To assess the mathematical content from other sections, you may use the problems from the Try This! section. In the Design-a-Test assessment, students may pick any goal or content from the unit.

Scoring

Answers are provided for all assessment problems. The method of scoring the problems depends on the types of questions in each assessment. Most questions require students to explain their reasoning or justify their answers. For these questions, the reasoning used by the students in solving the problems as well as the correctness of the answers should be considered as part of your grading scheme. A holistic scoring scheme can be used to evaluate an entire task. For example, after reviewing a student's work, you may assign a key word such as *emerging, developing, accomplishing,* or *exceeding* to describe his or her mathematical problem solving, reasoning, and communication.

DESIGN-A-TEST

Use additional paper as needed.

You will write a test for students your own age using the mathematics covered in this unit. This is your opportunity to show your teacher what you have learned.

To help you write the problems, you might look through newspapers and magazines for data sets and graphs or ask your family members about specific data and statistics they may use in their jobs. Try to vary the problems and questions you write. Some answers might be simple and straightforward, while others might require making a plot to analyze data or writing a conclusion based on data from a graph. You might also vary the difficulty of the problems.

Your test should also include solutions, point values, and scoring suggestions for the problems.

Be sure your work is completed neatly.

Good luck!

Solutions and Samples
of student work

The following are sample problems:

This is a list of the points scored by National Football League teams who played last Sunday: 27, 23, 20, 24, 14, 38, 13, 19, 16, 21, 10, 24, 34, 28, 36, 14, 20, 31.

1. Find the range, median, first quartile, and third quartile.

2. Make a box plot of these numbers.

3. What does your box plot tell you about the numbers?

Problems that attempt to address goals of reasoning, communication, and problem-solving for the same data set might be stated as follows:

1. Make a plot of the scores.

2. What information does this plot give you about the points scored by NFL football teams last Sunday?

3. Predict the typical score for an NFL game next Sunday.

Problem **1** in the second set of questions allows students to make their own decisions about an appropriate representation of the data.

Hints and Comments

Materials Design-a-Test assessment, page 157 of the Teacher Guide (one per student); graphing calculators, optional (one per student); graph paper, optional (one sheet per student); rulers, optional (one per student)

Overview Students design a test that reflects the mathematics in the unit. They also include solutions, point values, and scoring suggestions for their test problems.

About the Mathematics Encourage students to include problems that deal with the following goals: recognize possible bias in sample; determine whether representations of data (numerical and visual) are appropriate; and become aware of the questions that should be asked when analyzing data sets and representations of data.

Planning You may want students to work on this assessment activity individually or in small groups. Students' problems should cover concepts presented in the unit and include solutions, point values, and scoring suggestions. You may want to restrict students' problems to specific concepts. You may also want to place guidelines on other details, such as how graphs cut from printed resources are to be mounted and whether hand-drawn graphs should be on graph paper.

Scoring for the Design-a-Test assessment activity may be based on criteria that you feel are important. Students should know the criteria on which their work will be judged. You might discuss your scoring criteria when you assign the Design-a-Test assessment before the end of the unit. Scoring criteria might include the following:

- Do the questions ask for factual answers or for interpretation or decision making? (For example, some students might include only questions at the conceptual/procedural level, such as asking what a particular bar in a graph represents. Other students might ask questions that involve reasoning and communication, such as comparing two representations of data and suggesting ways to improve them.)
- Did the student include items requested when the assessment was assigned? (For example, did they include three different concepts, the answers, point values, and scoring suggestions?)
- Do the answers make sense?

HATS

Use additional paper as needed.

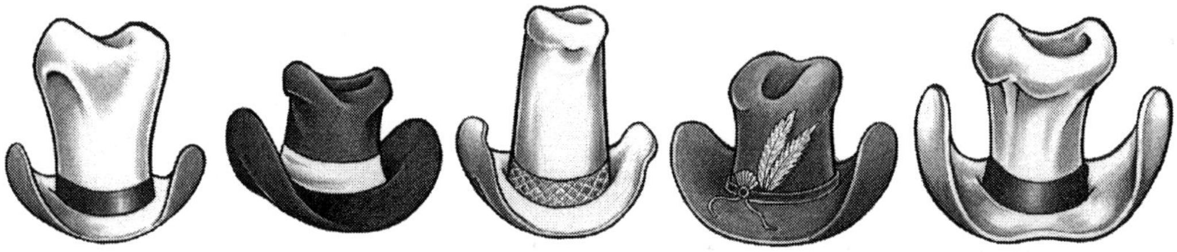

A company that sells hats wants to open a new shop featuring western-style hats. The company wonders what size hats to stock. The tables below list the heights and head circumferences in centimeters for 36 different adults.

Height (in cm)	174	190	176	183	180	182	193	182	168	172	189	163
Head Circumference (in cm)	55	58	57	60	60	58	58	59	57	56	59	58

Height (in cm)	168	190	198	167	186	169	174	163	175	180	176	175
Head Circumference (in cm)	57	61	59	57	60	55	57	56	60	62	52	54

Height (in cm)	178	184	196	185	174	185	171	179	172	185	180	176
Head Circumference (in cm)	50	62	60	58	58	59	56	61	56	57	56	60

1. What do you consider the typical head size? What does this number tell you?

The company has decided to stock the sizes that will fit the middle 50% of the adult population.

2. a. What kind of plot can you use to find the sizes within the middle 50% range?

b. What sizes will the company stock?

3. Investigate whether a relationship exists between head size and height. Is the result useful for the company?

4. What should the company find out before using the data to help them when ordering hats?

Solutions and Samples
of student work

1. Answers will vary. Sample responses:
   - The mode would be a good measure here. It tells you that 57 centimeters, 58 centimeters, and 60 centimeters are the most frequent head circumferences in the table. People who sell hats are interested in the mode (fitting the greatest number of customers).
   - The mean is 58 centimeters. It indicates the average head circumference in the table.
   - The median head size is 58 centimeters. This means that 58 is the point that separates the given set of head circumferences into two groups.

2. **a.** a box plot, because it indicates the spread of the middle 50% of the data

   b. Answers will vary. Sample response:

   The box plot indicates that 50% of the head circumferences are between 56 centimeters and 60 centimeters, so the company will stock sizes between 56 and 60 centimeters.

3. The scatter plot below indicates a weak but positive correlation. Answers about whether the result is useful to the company will vary. One possible answer is that, since the correlation is weak, the relationship between height and head circumference is not an issue the company will consider when ordering hats.

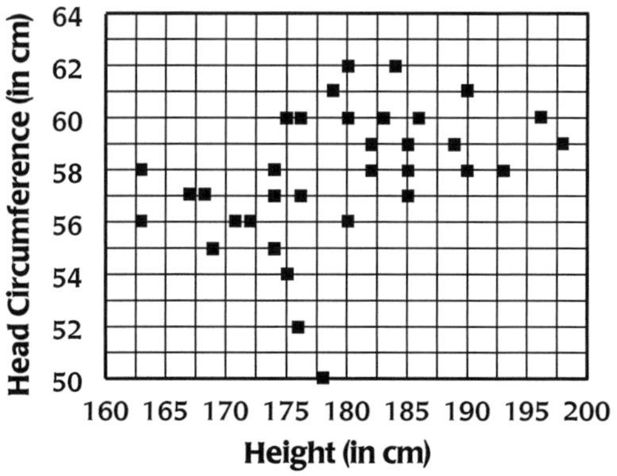

4. The company needs to consider whether bias resulted from the way the data were collected. The company should be aware that they are assuming the sample is representative of the population in the town in which the shop is located.

Hints and Comments

Materials Hats assessment, page 158 of the Teacher Guide (one per student); Student Activity Sheets 12 and 13, optional (one of each per student; graphing calculators, optional (one per student); graph paper, optional (one sheet per student); rulers, optional (one per student)

Overview Students investigate data on height and head circumference.

About the Mathematics In this assessment activity, students will show their ability to represent data graphically, describe data numerically, describe the relationship between two variables, identify the degree of correlation between variables, draw conclusions based on given data sets, and become aware of the questions that should be asked when analyzing data sets. For problem **4,** students must demonstrate a critical attitude toward the use of the presented data.

Planning You may want students to work on these assessment problems individually. *Note:* Student Activity Sheets 12 and 13 will help students who want to use a graphing calculator to solve these problems.

Comments about the Problems

1. Encourage students to explain their choices.

2. **a.** If students give good reasons for using another plot in which they can find the middle 50%, you might consider giving partial or full credit for the answer.

4. Encourage students to use what they have learned in the unit about sampling and looking critically at data.

MANATEES AND POWER BOATS

Use additional paper as needed.

Manatees are large mammals (10–13 feet long) that live in the waters off the coast of Florida. Manatees are in danger of extinction. In recent years, more and more manatees have been found dead on the beaches. The data below are numbers of dead manatees that have been recorded at different times.

13, 21, 24, 16, 24, 20, 15, 34, 33, 33, 39, 43, 50, 47

1. Look carefully at these data. Describe what the data tell you about the number of manatee deaths. You may want to use a plot of the data.

Because the bodies of the dead manatees had wounds on them that might have been caused by propellers from motor boats, a conservation group checked the number of power boats registered at the times the data were recorded. Then the group looked for a connection to the number of manatee deaths. The data they gathered are in the table on the right.

Power Boat Registrations (in thousands)	Manatee Deaths
447	13
460	21
481	24
498	16
513	24
512	20
526	15
559	34
585	33
614	33
645	39
675	43
711	50
719	47

Source: Decisions Through Data by David Moore, 1992 COMAP, Inc.

2. a. Make a scatter plot to show the relationship between the number of power boats registered and the number of manatee deaths.

b. Describe this relationship.

c. Describe the correlation in the scatter plot.

1. Answers will vary. Sample student responses:

   In general, the number of manatee deaths seems to be increasing over time.

   The range in the number of manatee deaths is between 13 and 50.

   The median is equal to 28.5.

   The mean is equal to 29.4.

   A box plot of the data:

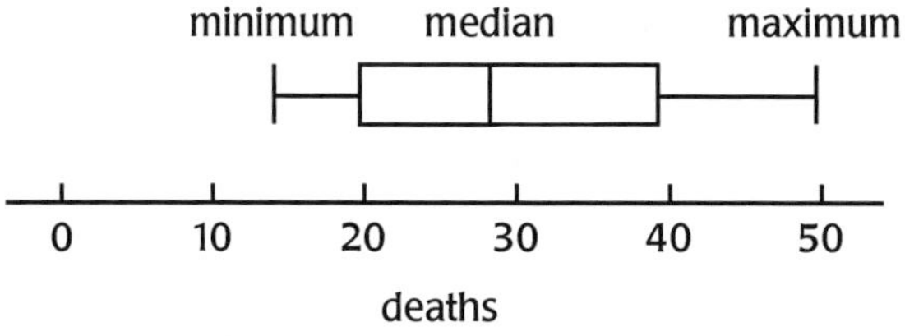

   A box plot shows that at least 50% of the data points fall between 20 and 39 deaths.

2. **a.**

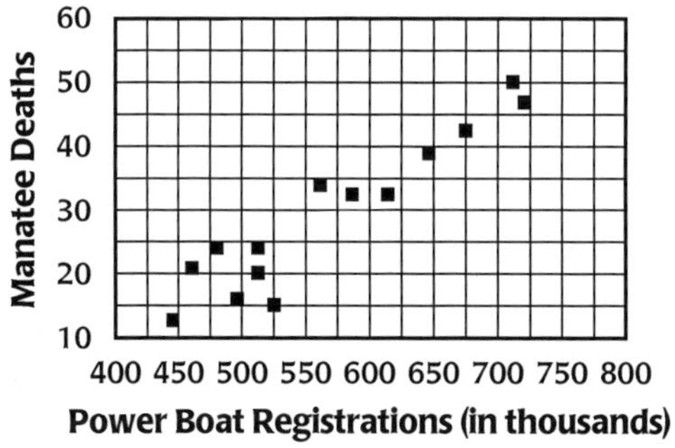

   b. The scatter plot shows that as the number of power boats increases, so does the number of manatee deaths.

   c. The correlation is positive and moderately strong.

Hints and Comments

Materials Manatees and Power Boats assessment, pages 159 and 160 of the Teacher Guide (one of each per student); graphing calculator, optional (one per student); graph paper, optional (one sheet per student); rulers, optional (one per student)

Overview Students investigate data on the deaths of manatees and show a relationship between manatee deaths and power boat registrations.

About the Mathematics In this assessment (on this and the next page), students need to demonstrate an ability to represent data graphically, describe data numerically, describe the relationship between two variables, identify the degree of correlation between variables, draw conclusions based on given data sets and representations of data, and become aware of the questions that should be asked when analyzing data sets and representations of data. In problem **4,** students need to demonstrate a critical attitude toward the use of the presented data.

Planning You may want students to work on these assessment problems individually.

Comments about the Problems

1. A plot over time cannot be constructed from the given data because no information on the time scale is provided. If students use a plot over time, they should be aware of the assumptions they are making.

   The disadvantage of using a range, median, mean, or box plot is that the order of the numbers is lost.

MANATEES AND POWER BOATS

Use additional paper as needed.

3. Estimate the number of manatee deaths for each number of power boats listed below. Explain how you determined your answers.

a. 450,000 power boats registered

b. 700,000 power boats registered

c. 600,000 power boats registered

4. What would you advise the Florida Department of Conservation to do about the problem of manatee deaths?

Solutions and Samples
of student work

3. a. Estimates will vary. Sample estimate:

About 14 manatees would be killed. Since 13 manatees were killed when there were 447 thousand power boat registrations, it seems likely that 14 or 15 manatees might be killed when slightly more power boats are registered.

b. Estimates will vary. Sample estimate:

About 45 manatees would be killed. Seven hundred thousand registrations is a little more than halfway between 675 and 711. An estimate might be a little less than halfway between the corresponding 43 and 50 manatee deaths.

c. Estimates will vary. Sample estimate:

About 33 manatees would be killed. Six hundred thousand registrations is about halfway between 588,000 and 614,000 registrations. An estimate might be halfway between the corresponding manatee deaths. Since both of these are 33, an estimate of 33 deaths is a good one.

4. Answers will vary. Sample student responses:

The Florida Department of Conservation could require that people who register their power boats take a course about manatees, their habits, the general areas in which they live, and the dangers of power boats to their safety.

The Florida Department of Conservation could recommend a restriction on the number of power boat registrations.

The Florida Department of Conservation could recommend banning power boats in areas in which manatees live.

Hints and Comments

Overview Students continue to investigate the relationship between manatee deaths and power boats and recommend solutions to the problem.

Planning You may want students to work on these assessment problems individually.

Comments about the Problems

3. Some students may estimate the number of manatee deaths from the scatter plot alone. Other students may draw a line of best fit, or graph a least squares linear regression line or a median fit line with their graphing calculators, to help them estimate the number of manatee deaths.

4. Students may want to write a letter to the Florida Department of Conservation explaining their study of the manatee deaths and include the statistics and graphs they used to solve problems **1–3.**

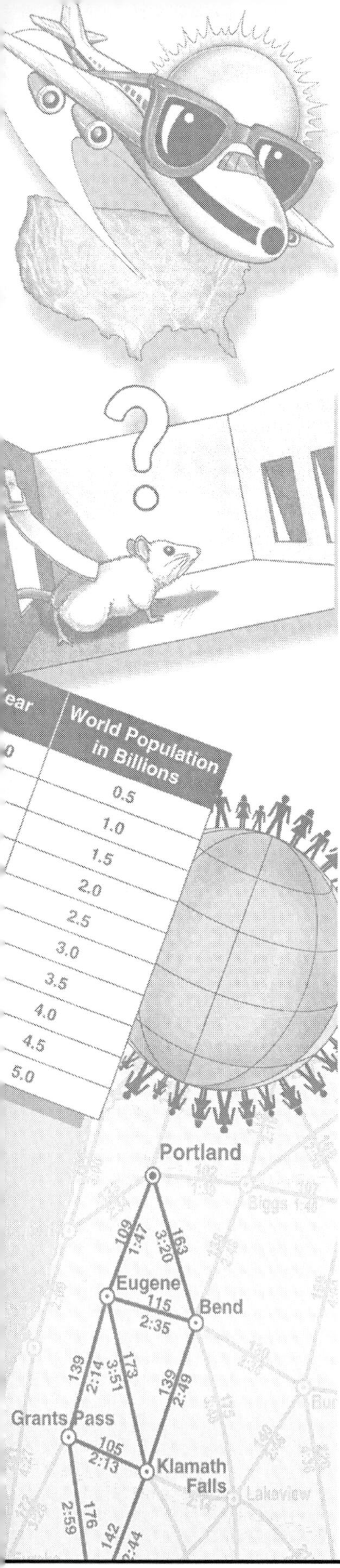

Insights Into Data

Glossary

bias (p. 38) an error introduced into sampling when elements have not been randomly selected; bias appears in sampling when certain outcomes are favored unfairly

box plot (p. 80) a graphic way of showing a summary of data using the median, quartiles, and extremes of the data; a box-and-whisker plot makes it easy to see how the data are spread out and where they are concentrated; the longer the box, the more spread out the data

cause-effect relationship (p. 98) a relationship in which a change in one variable directly causes a change in another variable

cluster (p. 20) a group of nearby points on a plot

correlation (p. 96) a relationship among variables

data point (p. 14) a point in a graph determined by an x-value and a corresponding y-value (the coordinates)

equation of a straight line (p. 46) $y = (\text{slope})x + (\text{intercept})$

least squares linear regression line (p. 112) a line used to predict data according to certain criteria

median fit line (p. 114) a line used to predict data according to certain criteria with random variations

origin (p. 52) a fixed point from which the x- and y-coordinates are measured

outlier (p. 20) a point that lies outside the cluster in a scatter plot

plot over time (p. 69) a graph that displays the change of a variable over time

population (p. 30) a complete set of outcomes or individuals that are of interest in a particular question

random number (p. 32) a number chosen by chance in order to eliminate bias due to personal choice

random sample (p. 38) a sample chosen by chance

sample (p. 32) a group taken from a larger population

scaling (p. 53) labeling the x-axis by marking a point 0 (origin), separating the line into equal parts, and labeling the separate points with the integers: 1, 2, 3 (on the right-hand side) or -1, -2, -3 (on the left-hand side); the y-axis is scaled in the same manner, but on a vertical line

scatter plot (p. 14) a graph that shows the relationship between two quantities, represented by points on the diagram

simulation (p. 32) a replicable experiment used to model some phenomenon

slope (p. 114) the steepness or the slant of a line; it is determined by the ratio of the coordinates of two points on a line

trial (p. 36) a sample used as part of a controlled experiment

y-intercept (p. 115) the point at which a graph intersects the y-axis

BRITANNICA Mathematics in Context

Blackline Masters

Letter to the Family

Dear Family,

Soon your child will begin the *Mathematics in Context* unit *Insights into Data.* Below is a letter to your child that opens the unit, describing the unit and its goals.

You can help your child relate the concepts in this unit to his or her own life by looking at newspaper or magazine articles that contain numerical data. Discuss how information is presented in many different ways. In this unit, students will learn to think critically about how numbers are used to justify claims or to draw conclusions. Your child will also learn how to make his or her own statements based on data.

Look for a survey in a newspaper or magazine and discuss it with your child. Find graphs in articles and have your child explain the messages conveyed by the graphs. Do you use graphs or surveys in your work? Your child might share your use of graphs or surveys with the rest of the class.

Enjoy working with your child to develop "insights into data."

Sincerely,

The Mathematics in Context Development Team

Dear Student,

Welcome to *Insights into Data.* Do you look at the graphs in newspapers to see if they make sense? Numbers and graphs are used as sources of data all around you: in sports, grades, sales and marketing, taxes, and car ratings.

Year	World Population (in billions)
1630	0.5
1820	1.0
1890	1.5
1930	2.0
1950	2.5
1960	3.0
1968	3.5
1975	4.0
1981	4.5
1987	5.0

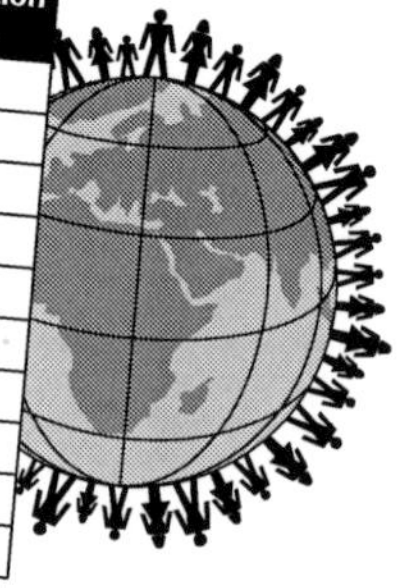

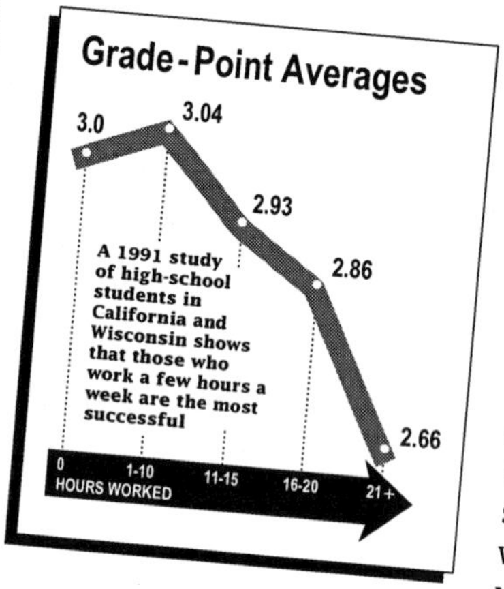

In this unit, you will learn how to use numbers and graphs to help you make decisions and draw conclusions. You will also study surveys and how they are conducted. You will grow mung beans in soda, salt water, and tap water to see which is the best solution for growing sprouts. You will even learn to use lines to help you investigate the relationship between two things, such as the height of a dinosaur and the length of its tail.

As you develop insights into data, keep your eye out for graphs in newspapers and magazines.

Sincerely,

The Mathematics in Context Development Team

Name________________________________

Student Activity Sheet 1

Use with *Insights into Data,* page 3.

Bean Number	Length of Bean Sprout (in millimeters)							Observations and/or Problems
	Day 1	Day 2	Day 3	Day 4	Day 5	Day 6	Day 7	
1								
2								
3								
4								
5								
6								
7								
8								
9								
10								

Student Activity Sheet 2

Name ____________________________________

Use with *Insights into Data,* pages 5–9.

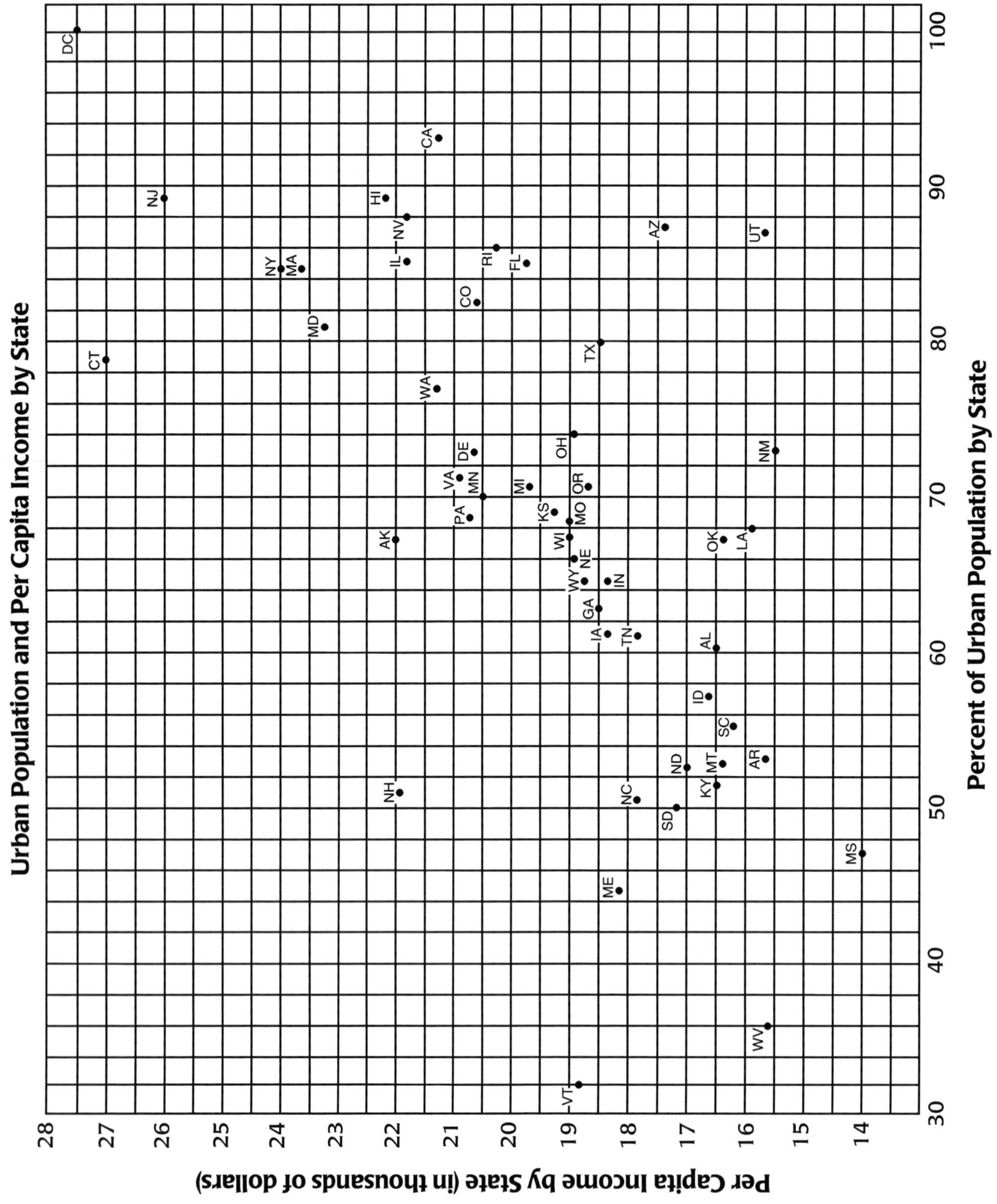

Source: Data from the U.S. Bureau of the Census, *Statistical Abstract of the United States: 1993,* (113th edition.) Washington, DC, 1993

Name________________________________

Student Activity Sheet 3

Use with *Insights into Data*, page 14.

Table of Random Numbers

39634	62349	74088	65564	16379	19713	39153	69459	17986	24537
14595	35050	40469	27478	44526	67331	93365	54526	22356	93208
30734	71571	83722	79712	25775	65178	07763	82928	31131	30196
64628	89126	91254	24090	25752	03091	39411	73146	06089	15630
42831	95113	43511	42082	15140	34733	68076	18292	69486	80468
80583	70361	41047	26792	78466	03395	17635	09697	82447	31405
00209	90404	99457	72570	42194	49043	24330	14939	09865	45906
05409	20830	01911	60767	55248	79253	12317	84120	77772	50103
95836	22530	91785	80210	34361	52228	33869	94332	83868	61672
65358	70469	87149	89509	72176	18103	55169	79954	72002	20582
72249	04037	36192	40221	14918	53437	60571	40995	55006	10694
41692	40581	93050	48734	34652	41577	04631	49184	39295	81776
61885	50796	96822	82002	07973	52925	75467	86013	98072	91942
48917	48129	48624	48248	91465	54898	61220	18721	67387	66575
88378	84299	12193	03785	49314	39761	99132	28775	45276	91816
77800	25734	09801	92087	02955	12872	89848	48579	06028	13827
24028	03405	01187	06316	81916	40170	53665	87202	88638	47121
86558	84750	43994	01760	69205	27937	45416	71964	52261	30781
78545	49201	05329	14182	10971	90472	44682	39304	19819	55799
14969	64623	82780	35686	30941	14622	04126	25498	95452	63937
58697	31973	06303	94202	62287	56164	79157	98375	24558	99241
38449	46438	91579	01907	72146	05764	22400	94490	49833	09258
62134	87244	73348	80114	78490	64735	31010	66975	28652	36166
72749	13347	65030	26128	49067	27904	49953	74674	94617	13317
81638	36566	42709	33717	59943	12027	46547	61303	46699	76243
46574	79670	10342	89543	75030	23428	29541	32501	89422	87474
11873	57196	32209	67663	07990	12288	59245	83638	23642	61715
13862	72778	09949	23096	01791	19472	14634	31690	36602	62943
08312	27886	82321	28666	72998	22514	51054	22940	31842	54245
11071	44430	94664	91294	35163	05494	32882	23904	41340	61185
82509	11842	86963	50307	07510	32545	90717	46856	86079	13769
07426	67341	80314	58910	93948	85738	69444	09370	58194	28207
57696	25592	91221	95386	15857	84645	89659	80535	93233	82798
08074	89810	48521	90740	02687	83117	74920	25954	99629	78978
20128	53721	01518	40699	20849	04710	38989	91322	56057	58573
00190	27157	83208	79446	92987	61357	38752	55424	94518	45205
23798	55425	32454	34611	39605	39981	74691	40836	30812	38563
85306	57995	68222	39055	43890	36956	84861	63624	04961	55439
99719	36036	74274	53901	34643	06157	89500	57514	93977	42403
95970	81452	48873	00784	58347	40269	11880	43395	28249	38743
56651	91460	92462	98566	72062	18556	55052	47614	80044	60015
71499	80220	35750	67337	47556	55272	55249	79100	34014	17037
66660	78443	47545	70736	65419	77489	70831	73237	14970	23129
35483	84563	79956	88618	54619	24853	59783	47537	88822	47227
09262	25041	57862	19203	86103	02800	23198	70639	43757	52064

Student Activity Sheet 4

Name ______________________________

Use with *Insights into Data,* page 15.

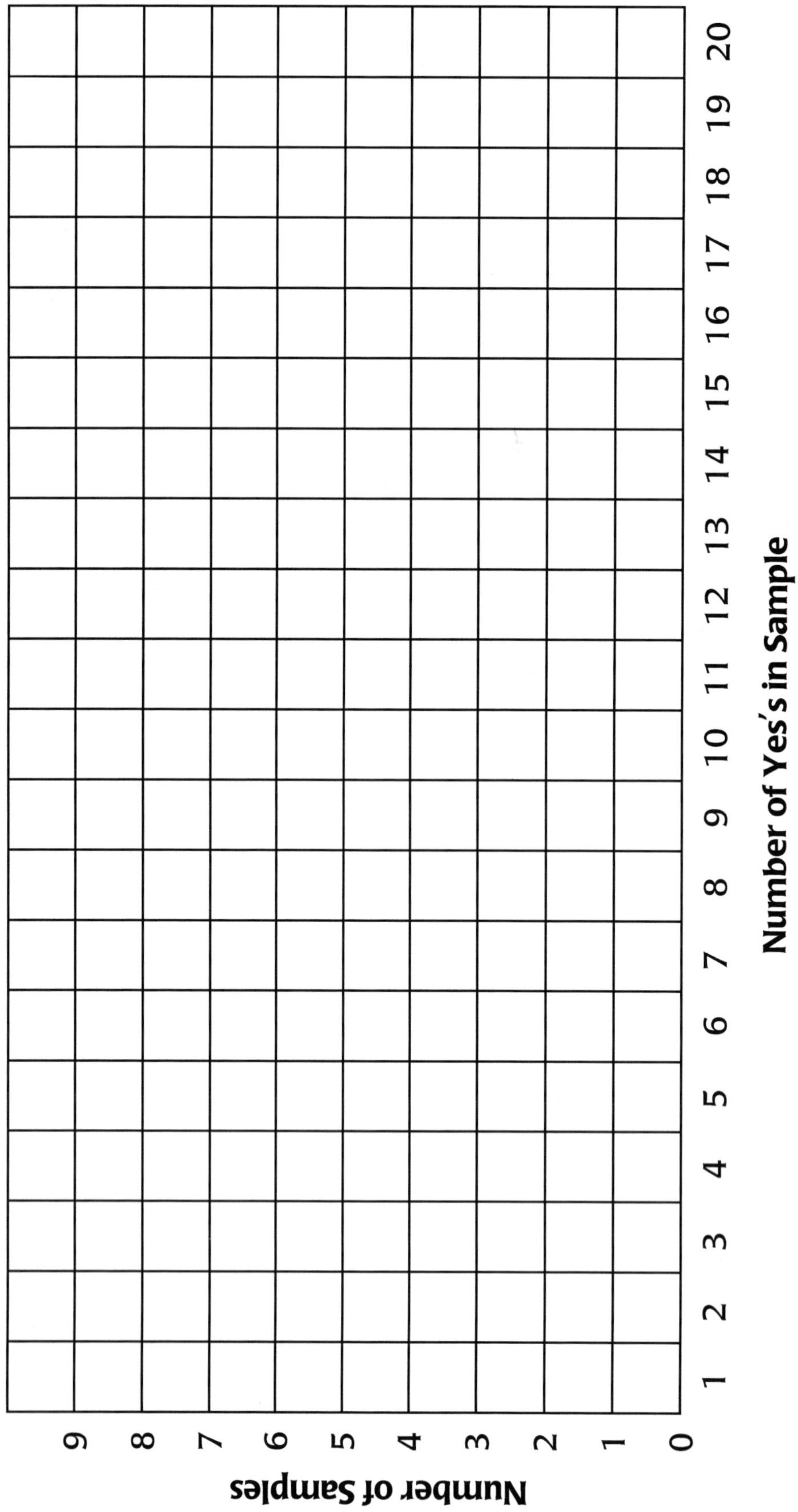

Name________________________________

Student Activity Sheet 5

Use with *Insights into Data*, page 23.

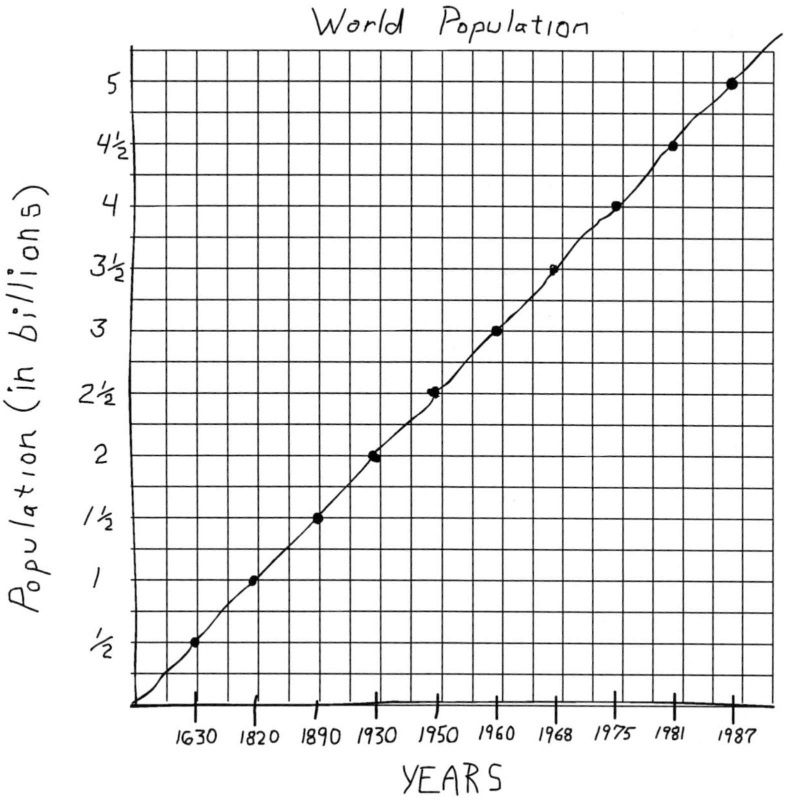

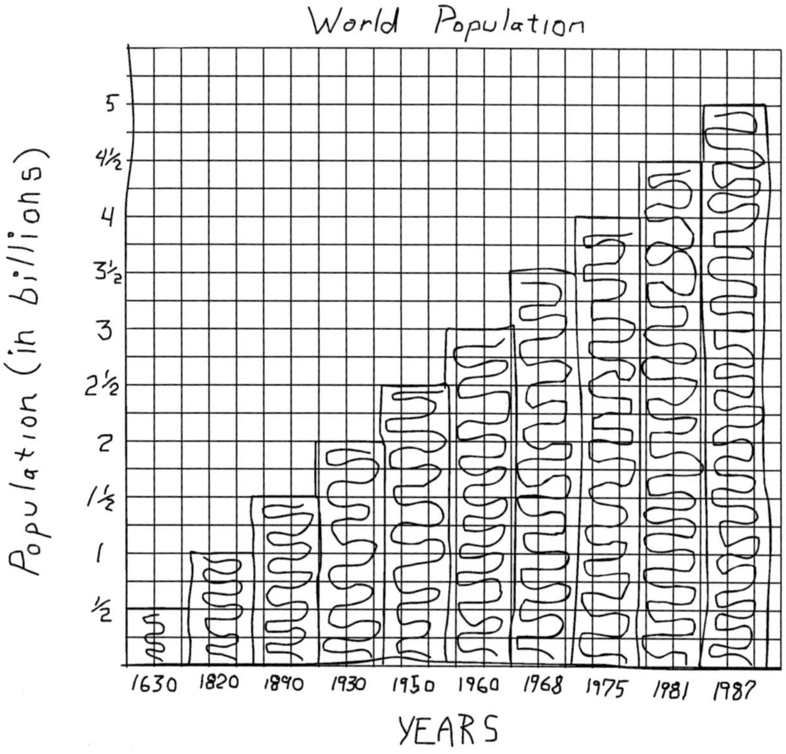

Student Activity Sheet 6

Name ______________________

Use with *Insights into Data*, pages 46 and 48.

Name of Bird	Bird Family	Nesting Grounds	Egg Length (in mm)	Egg Width (in mm)	Days to Hatch
Barn owl	Owl	Hollow trees, barns	39.2	30.8	30
Black cap	Warbler	Bushes, low trees	19.6	14.7	14-15
Black restart	Chat	Holes, crevices	19.4	14.4	13
Blackbird	Thrush	Trees, bushes	28.6	21.0	13-14
Blue titmouse	Titmouse	Holes	15.4	11.9	13-15
Brambling	Finch	Trees, bushes	19.5	14.6	12-13
Bullfinch	Finch	Bushes, conifers	20.2	15.1	13
Camion crow	Crow	Tall trees	43.5	30.1	17-19
Canada goose	Goose	Islets, marshes	85.7	58.2	28-30
Chaffinch	Finch	Bushes, trees, hedges	19.3	14.6	12-13
Chiff-chaff	Warbler	Wood verges, hedge rows	15.5	12.0	13-14
Chough	Crow	Crevices, cliff ledges	39.4	27.9	19-21
Cirl bunting	Bunting	Bushes,hedges	20.9	15.9	11-13
Coal titmouse	Titmouse	Holes, burrows	14.7	11.6	14-16
Collared dove	Pigeon	Trees, bushes	31.9	24.0	15-16
Common sandpiper	Sandpiper	Lake shores	36.3	36.0	21-22
Corn bunting	Bunting	Grassy hollows	23.8	17.7	12-13
Crested titmouse	Titmouse	Tree holes	16.0	12.4	15-18
Crossbill	Finch	Conifers	22.1	16.1	12-13
Dotterel	Plover	Moors, scree	41.1	28.9	27-28
Dunlin	Sandpiper	Moors, marshes	34.7	24.7	17
Eider	Duck	Near sea islets	77.6	51.9	25-26
Fieldfare	Thrush	Woodland trees	28.8	20.9	13-14
Firecrest	Warbler	Conifers	13.5	10.3	14-15
Gadwall	Duck	Reedbeds	51.8	37.5	26
Garden warbler	Warbler	Shrubs, brambles	20.1	14.8	12
Garganey	Duck	Dry ground, vegetation	45.3	33.3	21-23
Goldcrest	Warbler	Tall conifers	13.6	10.3	16
Golden plover	Plover	Boggy moors	51.8	35.9	27
Goldeneye	Duck	Tree holes, burrows	58.4	43.2	30
Goldfinch	Finch	Trees	17.0	12.8	12-13
Goosander	Duck	Trees, holes	66.4	46.4	32
Grasshopper warbler	Warbler	Moors, marshes	18.1	13.8	13-15
Great titmouse	Titmouse	Tree holes	17.3	13.5	13-14
Green sandpiper	Sandpiper	Trees on marshy ground	39.1	28.0	20-22

Name________________________________

Student Activity Sheet 7

Use with *Insights into Data*, pages 46 and 48.

Name of Bird	Bird Family	Nesting Grounds	Egg Length (in mm)	Egg Width (in mm)	Days to Hatch
Greenfinch	Finch	Bushes, hedges	20.2	14.5	13-14
Grey wagtail	Wagtail	Holes, hedges	19.0	14.5	12-14
Greylag goose	Goose	Reedbeds	85.3	58.0	28-29
Hawfinch	Finch	Deciduous trees	24.5	17.5	14
Hooded crow	Crow	Trees, bushes	43.5	30.1	17-19
House sparrow	Sparrow	Holes, crevices	22.5	15.7	13
Jackdaw	Crow	Holes, crevices	33.7	25.2	17-18
Jay	Crow	Bushes, trees	31.6	23.0	16-17
Lapwing	Plover	Fields, moors	47.1	33.7	24-26
Lesser whitethroat	Warbler	Thickets, hedge rows	16.5	12.6	10-11
Linnet	Finch	Bushes, thickets	17.7	13.3	12-14
Little owl	Owl	Tree holes	33.6	28.1	28
Little ringed plover	Plover	Mud, gravel pits	29.8	22.1	23-26
Long-eared owl	Owl	Abandoned nests	40.9	32.7	27-28
Long-tail titmouse	Titmouse	Trees, thickets	13.6	10.9	12-13
Magpie	Crow	Tall trees, hedges	34.1	24.2	17-18
Mallard	Duck	Ground cover	58.4	39.5	28
Marsh titmouse	Titmouse	Tree holes	16.1	12.2	14
Mistle thrush	Thrush	Tall trees	31.2	22.3	13-14
Nightingale	Chat	Ground cover	21.0	15.6	13
Pied wagtail	Wagtail	Holes, niches	20.4	15.1	12-14
Pintail	Duck	Moors, marshes	51.8	37.0	22-23
Pochard	Duck	Reeds, banks	61.3	43.7	24-26
Raven	Crow	Ledges, crevices, trees	49.7	33.4	20-21
Redpoil	Finch	Scrub growth	16.9	12.6	10-11
Redwing	Thrush	Bushes, trees	25.8	19.2	13
Reed bunting	Bunting	Reeds, marshes	19.3	14.3	13-14
Reed warbler	Warbler	Reeds, bushes	18.3	13.6	11-12
Restart	Chat	Tree holes	18.7	13.8	13-14
Ring ouzel	Thrush	Bushes, small trees	30.4	21.5	14
Ringed plover	Plover	Beaches, mud	35.7	25.9	24-27
Robin	Chat	Hollows, holes	19.4	14.8	13-14

Student Activity Sheet 8

Name ______________________________

Use with *Insights into Data*, pages 46 and 48.

Name of Bird	Bird Family	Nesting Grounds	Egg Length (in mm)	Egg Width (in mm)	Days to Hatch
Rock dove	Pigeon	Cliff holes	39.3	29.1	17
Rook	Crow	Tall trees	41.0	28.3	16-18
Scaup	Duck	Moors, islands	63.2	43.5	24-25
Sedge warbler	Warbler	Reeds, thickets	17.7	13.1	12-13
Shelduck	Duck	Sandy burrows	65.8	47.6	28-29
Short-eared owl	Owl	Moors, marshes	40.1	31.8	26-27
Shoveler	Duck	Moors, marshes	51.8	37.0	23-25
Siskin	Finch	Conifers	16.4	12.3	11-12
Skylark	Lark	Open ground	24.1	16.8	14
Snow bunting	Bunting	Stony ground	22.4	16.8	10-13
Song thrush	Thrush	Bushes, trees	27.3	20.4	12-13
Stock dove	Pigeon	Tree holes	37.9	29.0	16-17
Stonechat	Chat	Grassy hollows	18.9	14.4	14-15
Tawny owl	Owl	Hollow trees	48.2	38.7	28-29
Teal	Duck	Dry ground, vegetation	45.6	33.5	21-22
Temminck's stint	Sandpiper	Islets	28.0	20.4	13-14
Tree sparrow	Sparrow	Holes, crevices	19.3	14.0	13-14
Turtle dove	Pigeon	Trees, hedges	30.7	23.0	13-15
Twite	Finch	Ground cover	17.2	12.9	12-13
Wheatear	Chat	Holes, burrows	21.2	15.9	14
Whinchat	Chat	Grassy hollows	19.2	14.8	13-14
Whitethroat	Warbler	Scrub, bushes	18.1	13.8	11-13
Wigeon	Duck	Islets, ground cover	53.9	38.2	22-23
Wood sandpiper	Sandpiper	Boggy moors	38.3	27.0	21-24
Wood warbler	Warbler	Woodland ground	16.1	12.6	13
Wood-pigeon	Pigeon	Trees, bushes	40.1	28.7	15-17
Woodlark	Lark	Wood verges	21.6	16.3	13-15
Yellow wagtail	Wagtail	Grassy hollows	19.1	14.3	13-14
Yellowhammer	Bunting	Grassy hollows	21.2	15.9	12-14

Name________________________________

Student Activity Sheet 9

Use with *Insights into Data*, page 47.

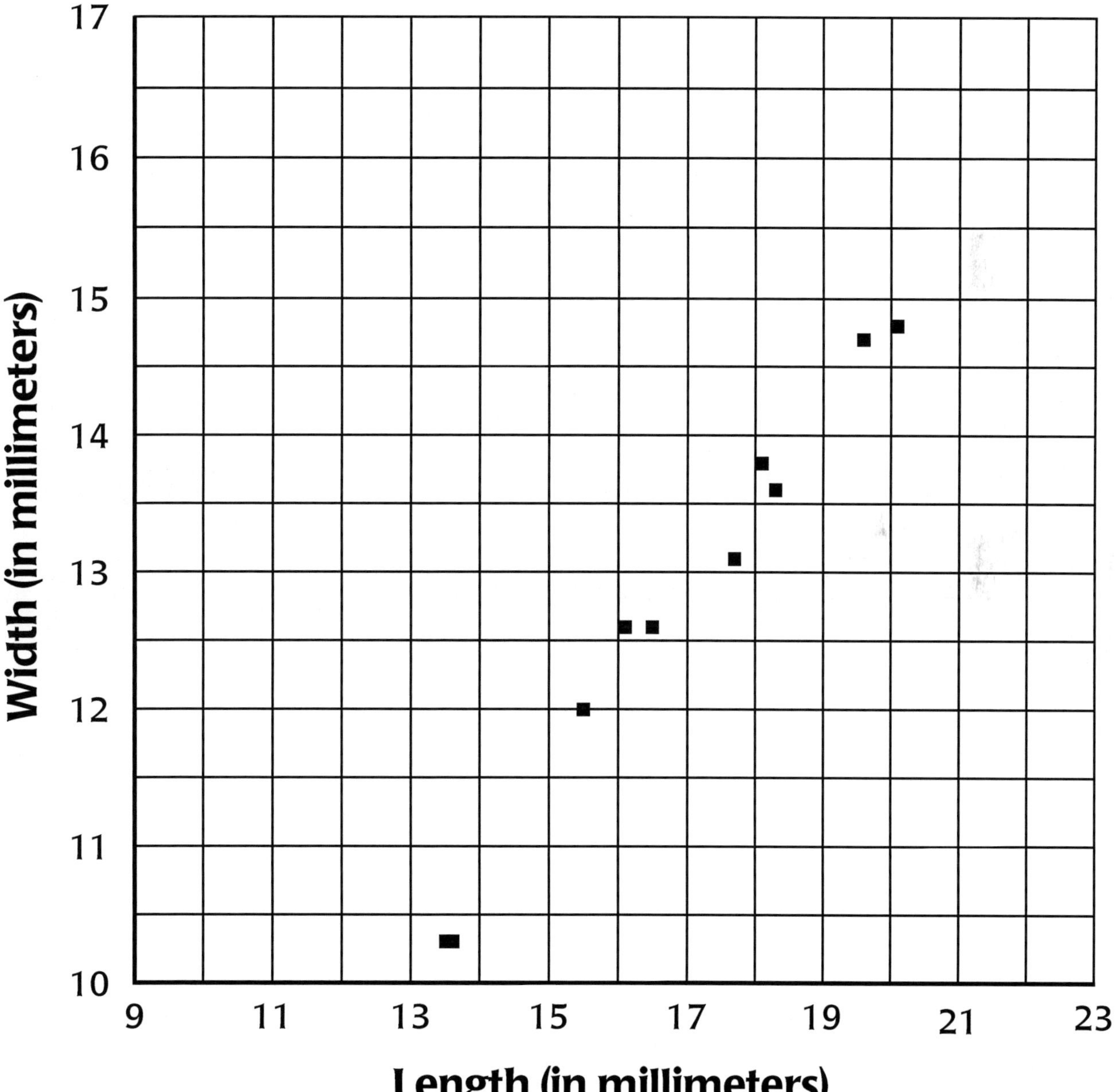

Student Activity Sheet 10

Name ______________________________

Use with *Insights into Data,* page 14.

Generating Random Numbers

I. Seed the Generator

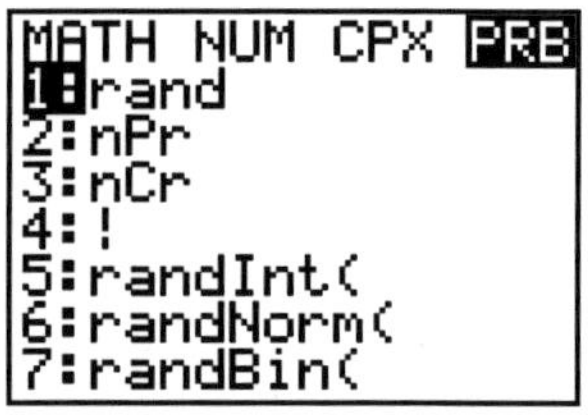

So that everyone in the class gets a different set of random numbers, each student should select a different seed value.

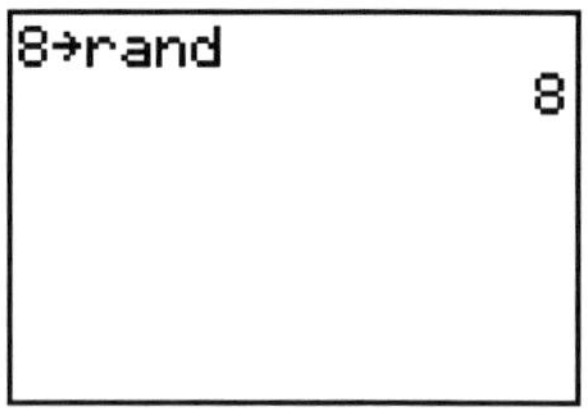

To set a seed value, store any integer as the random number generator. Type any integer into the calculator. Press STO► (RCL X). Then press MATH (TEST A). Arrow over to the **PRB** menu and then down to **1:rand.** Press ENTER. Press ENTER again to store the integer as the seed. The screen on the right shows 8 stored as the seed value.

II. Generate the Random Number

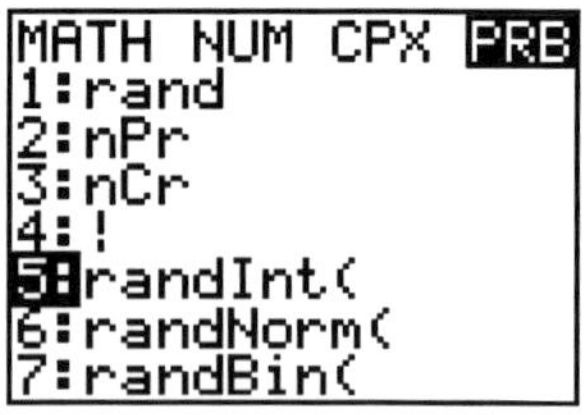

The “randInt” command on your calculator will generate random integers between any two numbers that you specify. For example, in order to generate random integers from 0 to 9, you need to input the expression “randInt(0,9)” on your calculator. Below, instructions are given on how to input this expression.

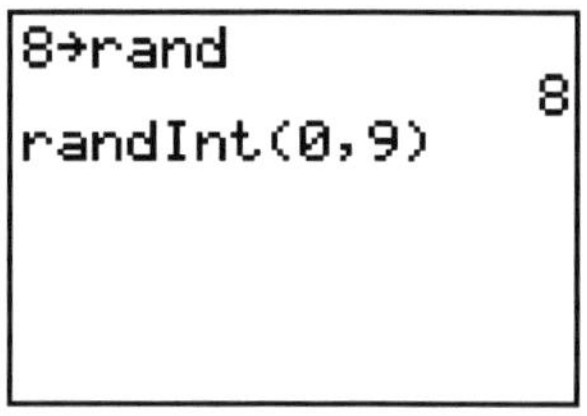

To generate a random integer from 0 to 9, press MATH (TEST A), arrow to **PRB,** and select **5:randInt(.** Then press 0 (CATALOG) , (EE J) 9 (W Q), and close the sequence by pressing) (} L). Your screen should look like the one on the right.

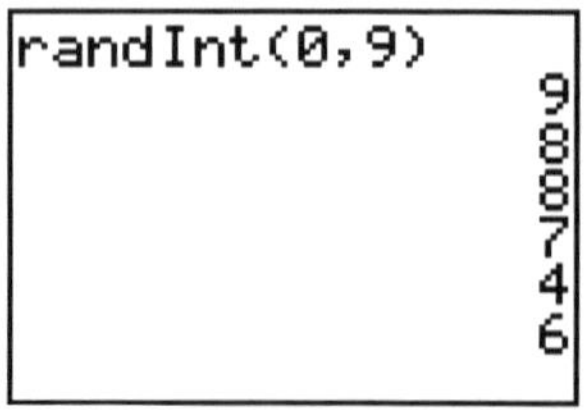

Press ENTER to get a random number.

Keep pressing ENTER to get more random numbers from 0 to 9.

Name________________________________

Student Activity Sheet 11

Use with *Insights into Data,* page 15.

Making a Histogram

I. Enter the Data

To reach the data tables, press LIST [STAT] and choose **1:Edit** (press [ENTER] or the number **1**). If there are numbers in the list, you must delete the entries. One way to do this is to use the arrow keys and highlight the List name (L1) and press [CLEAR] and [ENTER]. Then enter your data one value at a time under L1, pressing [ENTER] after each value entered.

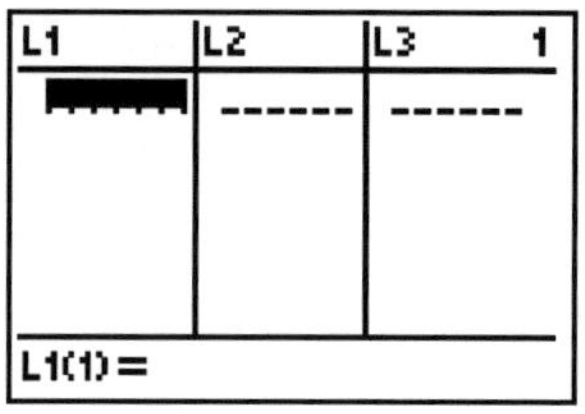

II. Set up a Viewing Window

Inspect your data and decide what range of values you are working with. The lowest data point should be your **Xmin** and a value a little larger than your highest data point should be your **Xmax.**

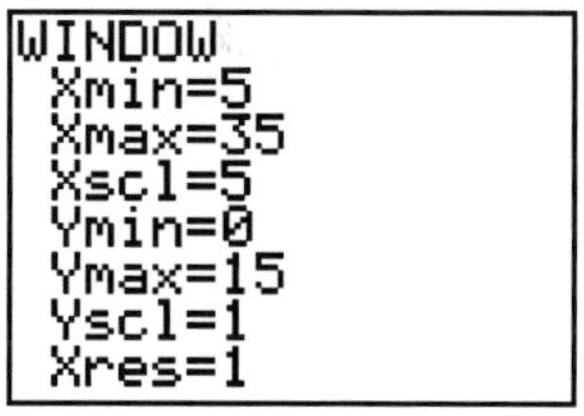

To set your viewing window, press TBLSET [WINDOW]. Arrow down to enter the maximum and minimum data values beside **Xmin** and **Xmax.** Enter the width for the bars of the histogram under **Xscl.** A good value for **Ymin** is 0. Under **Ymax** enter a value that is larger than the largest frequency in any interval. **Yscl** is the distance between the scale marks along the vertical axis.

III. Make the Histogram:

Before you make your histogram, make sure that all the unwanted plots are turned off. Select the **STAT PLOT** menu by pressing [2nd] STAT PLOT [Y=]. One way to turn off the plots is to highlight **OFF** in each individual plot window. Or select **4:PlotsOff** and press [ENTER]. When you have finished, "Done" should appear on the home screen.

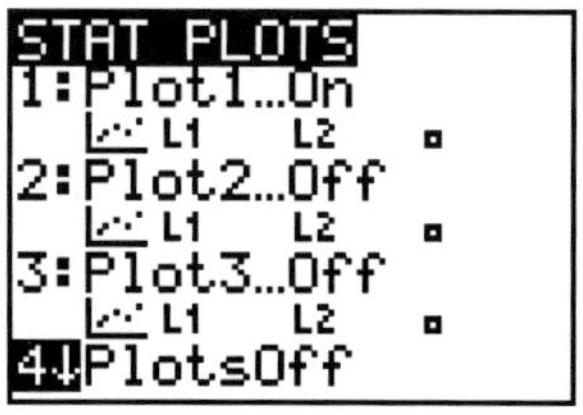

To choose the histogram, press [2nd] STAT PLOT [Y=] and choose **1:Plot 1.** Then make the selections as shown in the window on the right. When finished, press [2nd] QUIT [MODE]. Press TABLE [GRAPH] to see the histogram. Use the CALC [TRACE] key and the left and right arrows to investigate the points. The values on the screen give the minimum and maximum value for each interval, and the frequency of the interval.

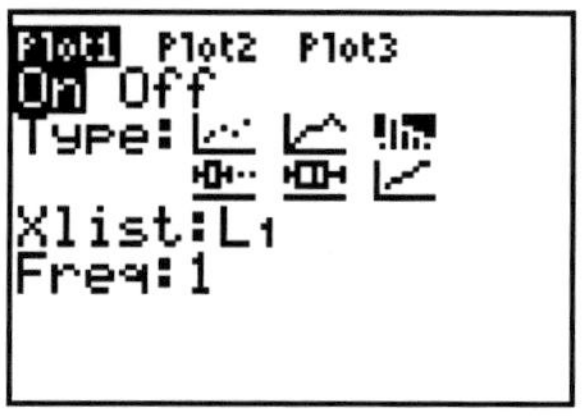

Student Activity Sheet 12

Name ________________________________

Use with *Insights into Data,* page 19.

Making a Box Plot

I. Enter the Data

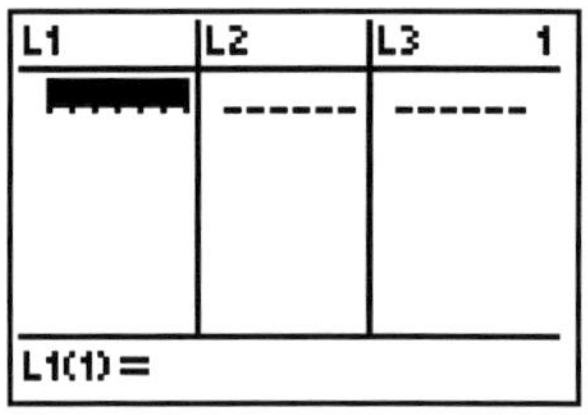

To reach the data tables, press LIST [STAT] and choose **1:Edit** (press [ENTER] or the number **1**). If there are numbers in the list, you must delete the entries. One way to do this is to use the arrow keys and highlight the List name (L1) and press [CLEAR] and [ENTER]. Then enter your data one value at a time under L1, pressing [ENTER] after each value entered.

II. Set up a Viewing Window

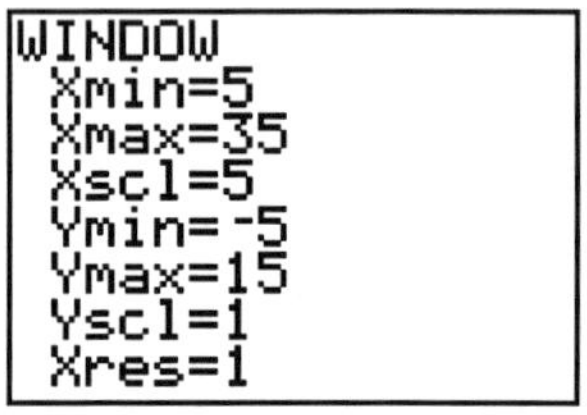

Inspect your data and decide what range of values you are working with. The lowest data point should be your **Xmin** and a value a little larger than the highest data point should be your **Xmax.** To reach the viewing window, press TBLSET [WINDOW]. Then enter the maximum and minimum data values for **Xmin** and **Xmax.** Any **Xscal** and **Y** values work.

III. Make a Box Plot

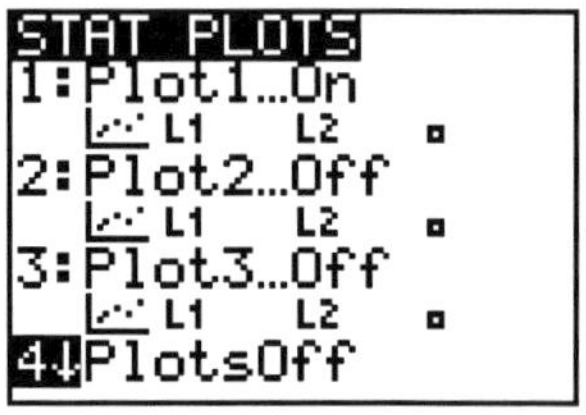

Before you make your box plot, make sure that all the unwanted plots are off. Select the **STAT PLOT** menu by pressing [2nd] STAT PLOT [Y=]. One way to turn off the plot is to select **4:PlotsOff** and press [ENTER]. When you have finished, "Done" should appear on the home screen.

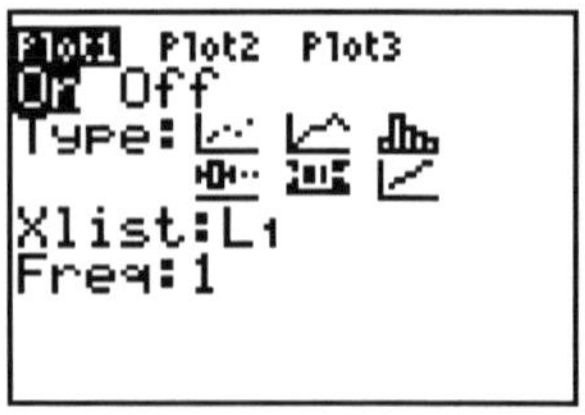

To choose the box plot, press [2nd] STAT PLOT [Y=] and choose **1:Plot 1.** Then make the selections as shown in the screen on the right. These selections make a box plot of the data in L1. When finished, press [2nd] QUIT [MODE].

Press TABLE [GRAPH] to see the box plot. Use the CALC [TRACE] key and the left and right arrows to investigate the quartile values.

Name________________________________

Student Activity Sheet 13

Use with *Insights into Data,* page 46.

Making a Scatter Plot

I. Enter the Data

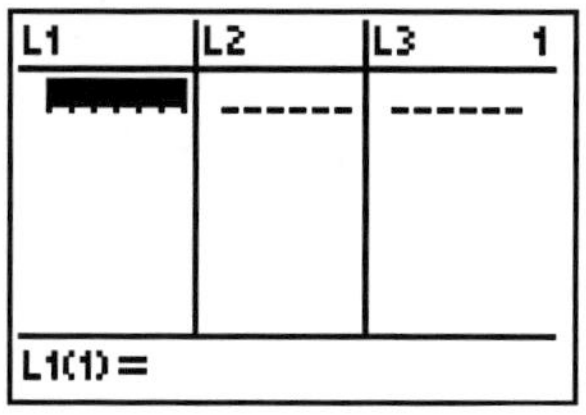

To reach the data tables, press LIST [STAT] and choose **1:Edit** (press [ENTER] or the number **1**). If there are numbers in lists L1 or L2, you must delete the entries. One way to do this is to use the arrow keys and highlight the List name (L1 or L2) and press [CLEAR] and [ENTER]. Then enter your data for your first list one value at a time under L1, pressing [ENTER] after each value entered. When you have finished, enter your data for your second list under L2.

II. Set up a Viewing Window

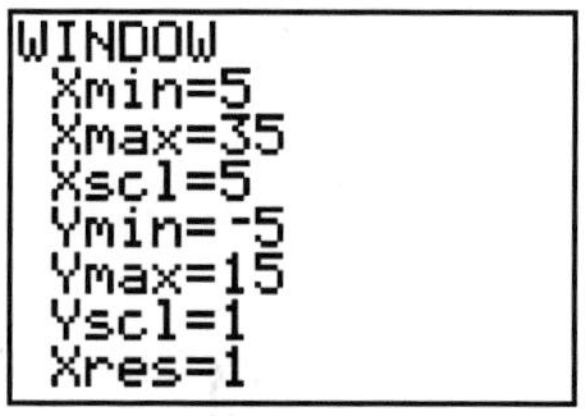

To choose an appropriate viewing window, press TBLSET [WINDOW]. Then enter the minimum and maximum L1 values for **Xmin** and **Xmax,** and the minimum and maximum L2 values for **Ymin** and **Ymax.** The **Xscl** is the scale along the x-axis and **Yscl** is the scale along the y-axis. Select appropriate values for each.

III. Make a Scatter Plot

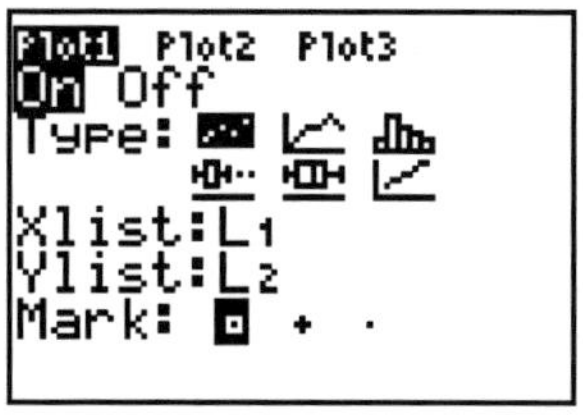

Before you make your scatter plot, make sure that all unwanted plots are off. To choose the scatter plot, press [2nd] STAT PLOT [Y=] and choose **1:Plot 1.** Then make the selections as shown in the window on the right. When finished, press [2nd] QUIT [MODE].

Press TABLE [GRAPH] to see the scatter plot. Use the CALC [TRACE] key and the left and right arrows to investigate the points.

Student Activity Sheet 14

Name ______________________________

Use with *Insights into Data,* pages 49 and 50.

Fitting Your Data with a Line

I. Make a Scatter Plot

See **Student Activity Sheet 13.**

II. Graph a Line That Fits Your Data

To graph the line that best fits your data, you will make a line called the least squares linear regression line.

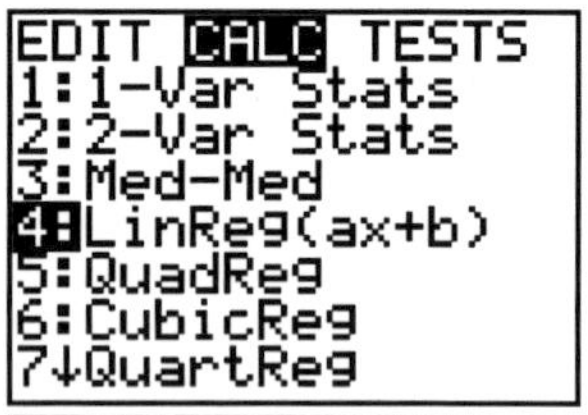

Press [LIST / STAT] and arrow over to the **CALC** menu.

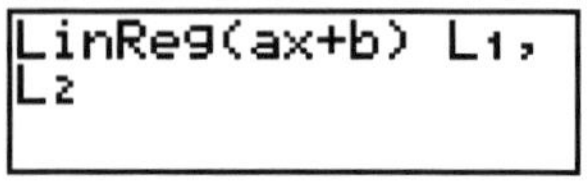

Choose **4:LinReg(ax+b)** and enter **L1, L2** by pressing [2nd] [L1 Y / 1] [EE J / ,] [2nd] [L2 Z / 2].

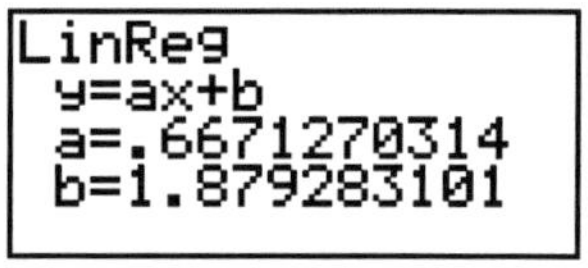

Then press [ENTER]. Your screen will display information for the line that fits your data. A sample screen is shown to the right.

III. Store the Line and Graph It

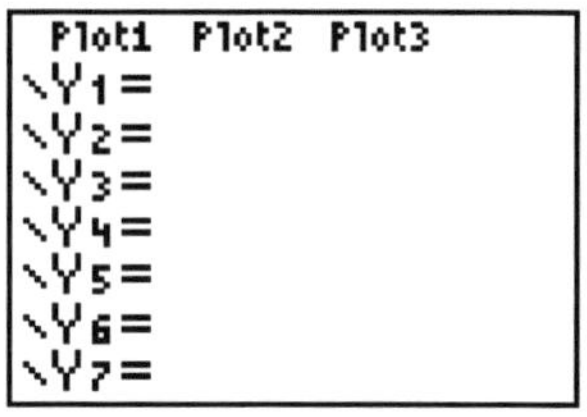

First you need to put the regression equation into your list of equations. Press [STAT PLOT / Y=]. Erase any previously entered equations by arrowing down to the equation and pressing [CLEAR].

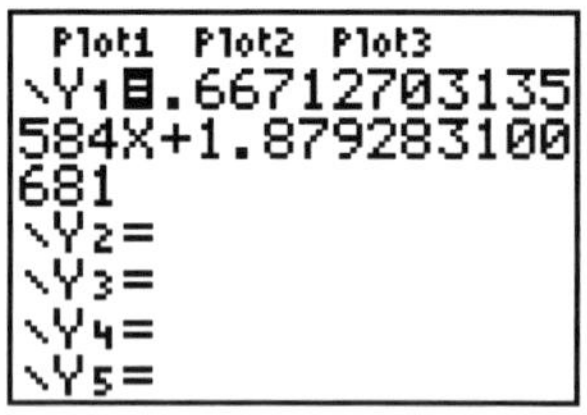

Press [DISTR / VARS] and choose **5:Statistics.** Arrow over to the **EQ** menu, and then choose **1:RegEq.** The equation should appear back in your **Y=** list of equations. The screen on the right shows that the regression equation is now stored in Y_1.

Press [TABLE / GRAPH] to see your equation and scatter plot.

IV. Extension

To graph a different line (called the median fit line), in Step II above, choose **3:Med-Med** instead of **4:LineReg(ax+b)** and then continue with the steps provided. There is actually no end to the ways to fit a set of data with an equation. Line fitting is extremely important in many mathematical applications.

Name ______________________ Date ______________________

DESIGN-A-TEST

Use additional paper as needed.

You will write a test for students your own age using the mathematics covered in this unit. This is your opportunity to show your teacher what you have learned.

To help you write the problems, you might look through newspapers and magazines for data sets and graphs or ask your family members about specific data and statistics they may use in their jobs. Try to vary the problems and questions you write. Some answers might be simple and straightforward, while others might require making a plot to analyze data or writing a conclusion based on data from a graph. You might also vary the difficulty of the problems.

Your test should also include solutions, point values, and scoring suggestions for the problems.

Be sure your work is completed neatly.

Good luck!

Name ______________________ Date ______________________

HATS

Use additional paper as needed.

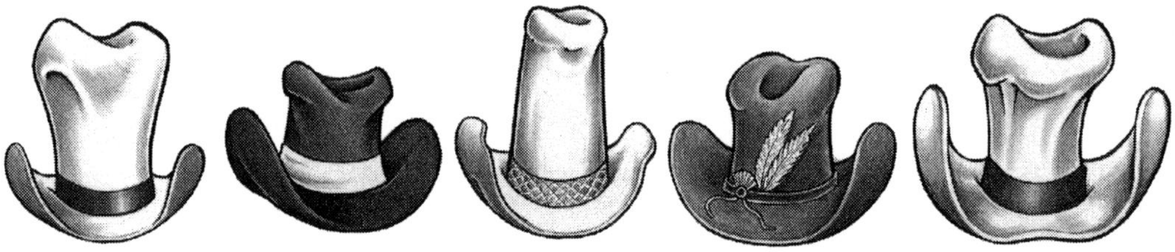

A company that sells hats wants to open a new shop featuring western-style hats. The company wonders what size hats to stock. The tables below list the heights and head circumferences in centimeters for 36 different adults.

Height (in cm)	174	190	176	183	180	182	193	182	168	172	189	163
Head Circumference (in cm)	55	58	57	60	60	58	58	59	57	56	59	58

Height (in cm)	168	190	198	167	186	169	174	163	175	180	176	175
Head Circumference (in cm)	57	61	59	57	60	55	57	56	60	62	52	54

Height (in cm)	178	184	196	185	174	185	171	179	172	185	180	176
Head Circumference (in cm)	50	62	60	58	58	59	56	61	56	57	56	60

1. What do you consider the typical head size? What does this number tell you?

The company has decided to stock the sizes that will fit the middle 50% of the adult population.

2. a. What kind of plot can you use to find the sizes within the middle 50% range?

b. What sizes will the company stock?

3. Investigate whether a relationship exists between head size and height. Is the result useful for the company?

4. What should the company find out before using the data to help them when ordering hats?

Name ______________________ Date ______________________

MANATEES AND POWER BOATS

Use additional paper as needed.

Manatees are large mammals (10–13 feet long) that live in the waters off the coast of Florida. Manatees are in danger of extinction. In recent years, more and more manatees have been found dead on the beaches. The data below are numbers of dead manatees that have been recorded at different times.

13, 21, 24, 16, 24, 20, 15, 34, 33, 33, 39, 43, 50, 47

1. Look carefully at these data. Describe what the data tell you about the number of manatee deaths. You may want to use a plot of the data.

Because the bodies of the dead manatees had wounds on them that might have been caused by propellers from motor boats, a conservation group checked the number of power boats registered at the times the data were recorded. Then the group looked for a connection to the number of manatee deaths. The data they gathered are in the table on the right.

2. a. Make a scatter plot to show the relationship between the number of power boats registered and the number of manatee deaths.

b. Describe this relationship.

c. Describe the correlation in the scatter plot.

Power Boat Registrations (in thousands)	Manatee Deaths
447	13
460	21
481	24
498	16
513	24
512	20
526	15
559	34
585	33
614	33
645	39
675	43
711	50
719	47

Source: Decisions Through Data by David Moore, 1992 COMAP, Inc.

Name ______________________________ Date ______________________________

MANATEES AND POWER BOATS

Use additional paper as needed.

3. Estimate the number of manatee deaths for each number of power boats listed below. Explain how you determined your answers.

a. 450,000 power boats registered

b. 700,000 power boats registered

c. 600,000 power boats registered

4. What would you advise the Florida Department of Conservation to do about the problem of manatee deaths?

Section A. Patterns in Data

1. 2.0

2. Answers and explanations will vary. Sample responses:

a. Yes. One student who did not study at all had the lowest grade point average.

b. No. This was not always true. Five students who studied at least nine hours a week had grade point averages under 3.5.

3. Answers will vary. Sample response:

No. The points are all fairly close together.

Yes. One student studied 11 hours a week, but still had a grade point average of 2.0.

Possible outliers are circled below:

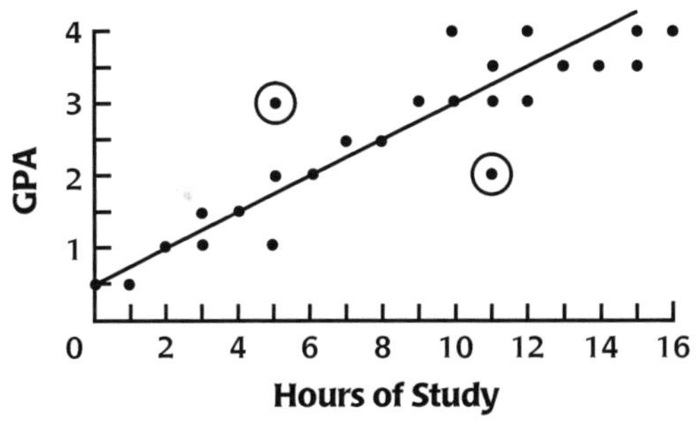

Section B. Selecting Samples

1. Answers and explanations will vary. Sample response:

   No, the statement is not reliable. Only two brands were tested, but Bingo Pop claims to be better than all leading brands.

2. Answers and explanations will vary. Sample response:

   No, this sample may include even more Bingo Pop drinkers, and those persons who don't like Bingo Pop or who were in a hurry may have avoided the booth.

3. Answers will vary. Sample responses:

   - Include more brands for testing.
   - Move the taste test to a new location the second time.
   - Don't have all the testing at only one location.

Section C. Interpreting Graphs

1. Answers and explanations will vary. Sample responses:

   - No. The picture of the 1985 candy bar is far more than twice the size of the 1970 candy bar.
   - No. Using a picture to represent the cost of a candy bar is misleading. It appears as though in 1985 you get a candy bar that is four times the size but only twice the cost as in 1970.

2. Graphs will vary. Sample graph:

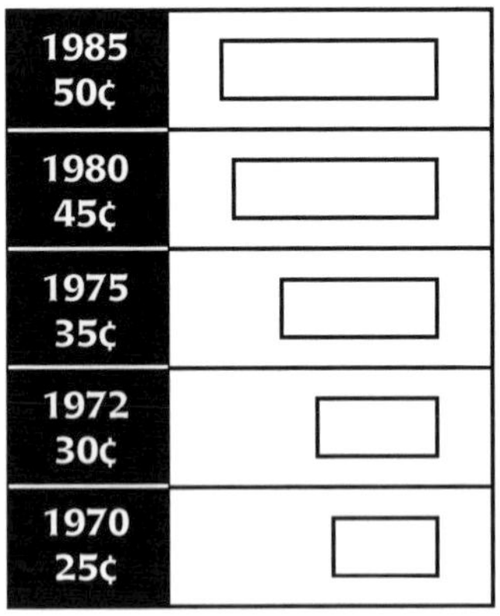

Section D. Using Plant Growth Data

1. a. Explanations will vary. Sample explanations:

Histogram: shows the distribution of values in a sample and can be used to estimate the mean; a two-dimensional histogram can only represent the distribution of one variable, such as GPA.

Box plot: shows the median and the middle 50% of the data; cannot find patterns based upon the distribution of values and cannot represent more than one variable.

Scatter plot: shows possible patterns or correlation between two variables; no average value can be taken from the plot.

b. Answers will vary. Sample response:

A scatter plot would be best because it can represent both variables in the same graph. Different colors can be used for the different classes.

2.

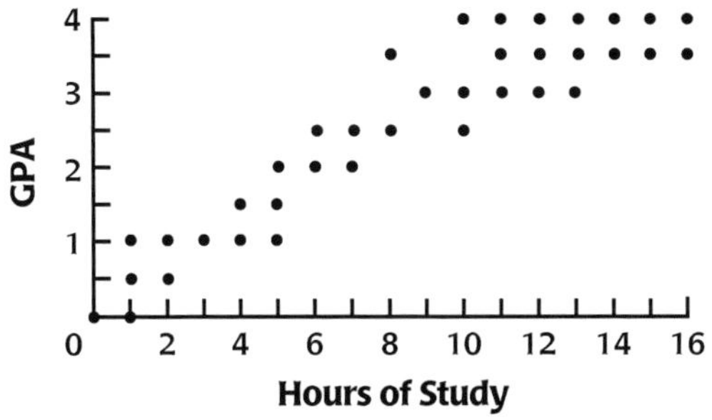

3. Answers will vary. Some students may say that the graph is evidence that those students who studied 11 or more hours per week had the highest grades (GPA = 3.0 or higher).

Section E. Correlating Data

1. a. weak correlation (not linear)

b. strong negative correlation

c. strong correlation (y-value is constant)

2. Answers will vary. Sample responses:

On each graph, if x is the price of an item and y represents the sales of that item:

a. Demand for some items may fall as the price increases, but at some point these items become so valuable that demand goes back up. Antiques are like this.

b. In general, when the price is low, sales are high. When the price increases, sales decrease.

c. Some necessary items like food and water may be sold at any price because people always need a certain amount.

3. Answers will vary. Sample response:

There are many examples from folklore: The groundhog does not affect the length of winter based upon whether or not it sees its shadow on February 2nd.

Section F. Lines That Summarize Data

1. a. Answers will vary. Some students may use the first point and the last point to calculate the vertical change and the horizontal change. The vertical change is 3.5 and the horizontal change is 16. The slope is $\frac{3.5}{16}$ or $\frac{7}{32}$.

b. Answers will vary. Sample response:

Using a ratio table:

÷ 7

7	1
32	$\frac{32}{7}$

$\frac{32}{7}$ is equal to about 4.5. So, for every 4.5 hours of studying, you may expect your grade point average to increase one point.

2. 2.5

3. a. Answers will vary. Sample responses:

If there are no outliers, can we minimize the distance from the line to each point?

If there are outliers, can we find a line in which one-half of the points are above the line and one-half of the points are below the line?

b. No. There are too many points above the line.

4. The median fit line should be used over the least squares linear regression line when there are outliers present.

Cover

Design by Ralph Paquet/Encyclopædia Britannica Educational Corporation.

Collage by Koorosh Jamalpur/KJ Graphics.

Title Page

Tom Zielinski/Encyclopædia Britannica Educational Corporation.

Illustrations

6 Phil Geib and Tom Zielinski/Encycloædia Britannica Educational Corporation; **8** Phil Geib/ Encyclopædia Britannica Educational Corporation; **10** Paul Tucker/ Encyclopædia Britannica Educational Corporation; **16, 18 (top, bottom)** Tom Zielinski/ Encyclopædia Britannica Educational Corporation; **18 (middle), 30, 32** Phil Geib/ Encyclopædia Britannica Educational Corporation; **34** Tom Zielinski/Encyclopædia Britannica Educational Corporation; **38** Paul Tucker/Encyclopædia Britannica Educational Corporation; **40, 42, 68, 70** Phil Geib/Encyclopædia Britannica Educational Corporation; **72 (bottom)** Tom Zielinski/Encyclopædia Britannica Educational Corporation; **92, 94, 104, 106** Phil Geib/Encyclopædia Britannica Educational Corporation; **108** Paul Tucker/Encyclopædia Britannica Educational Corporation; **112** Tom Zielinski/Encyclopædia Britannica Educational Corporation; **118 (top)** Paul Tucker/Encyclopædia Britannica Educational Corporation; **118 (bottom), 122** Phil Geib/Encyclopædia Britannica Educational Corporation; **132** Brent Cardillo/Encyclopædia Britannica Educational Corporation; **134** Phil Geib/Encyclopædia Britannica Educational Corporation.

Photographs

6 © Brent Cardillo/Encyclopædia Britannica Educational Corporation; **12, 28** © Cameramann International; **78, 80** Brent Cardillo/Encyclopædia Britannica Educational Corporation; **86, 88 (top), 90** © Encyclopædia Britannica Educational Corporation; **88 (bottom)** © Dave Wood/Encyclopædia Britannica Educational Corporation; **108** © This image was obtained from the Corel Professional Photos CD-ROM™; **110, 114** © Encyclopædia Britannica Educational Corporation; **120, 136, 138** © These images were obtained from the Corel Professional Photos CD-ROM™.